QUICK AND AMAZING CROSSWORD PUZZLE FOR GIRLS

The SuperGirl Edition (with 100 Drills!)

PUZZLE THERAPIST

CROSSWORD | SUDOKU | KIDS & ADULTS

CONTENTS

PUZZLE 1

ACROSS

1. Helgenberger of "CSI"

5. Virtual meeting of a sort

10. The English translation for the french word: kyste

14. Plymouth Reliant, for one

15. What 'check' could mean

16. With 10-Down, bun protectors

17. Requires more than one person, in a saying

20. Ali moniker, with "the"

21. Honolulu's ___ Palace

22. Weight not charged for

23. Mae West role

24. What Nashville sunbathers acquire?

29. #39 in a series

30. Uncertainties

31. Sunshine Biscuits brand

33. Underground org.

34. Lana Del ___, singer with the 2014 #1 album 'Ultraviolence'

35. Fill in the blank with this word: ""Roll ___ bones!""

36. Vent, in a way

38. Vandal

39. Turn red, perhaps

42. Event first won by a Marmon Wasp

45. Midnight alarm giver

46. Large food tunas

47. Setting of 'Beau Geste'

50. Hank Williams or Nat King Cole

54. Nursery rhyme fellow

56. It's you ___'

57. Tableware inspired by Scandinavian design

58. Waiting area announcements, briefly

59. Stand for the deceased

60. Small paving stones

61. Fixed at an acute angle

DOWN

1. Business school subj.

2. Rent-___

3. Fill in the blank with this word: "___ over the coals"

4. Social reformer Margaret Fuller, to Buckminster Fuller

5. One on the way in

6. Uses a divining rod

7. Literally, "father"

8. Fill in the blank with this word: "___ Offensive"

9. Study for astronomes

10. often formed into braided loaves and glazed with eggs before baking

11. Indian of the Sacramento River valley

12. Unscramble this word: sgin

13. U.S.S. Enterprise counselor

18. Batting coach's concern

19. Hop ___!'

23. Thumb one's nose at

24. Terrell who sang with Marvin Gaye

25. Gen. Rommel

26. Tuscan city

27. Nest

28. Dig discovery: Var.

29. You need one to break

32. Kipling's 'Follow Me ___'

34. Old character

37. First-rate

38. Five Norwegian kings

39. Nontraditional haircuts

40. Way out

41. Medicinal teas

43. Yesterday, in Italy

44. Most likely to sunburn

47. Mop: Var.

48. Gas: Prefix

49. The English translation for the french word: hËme

50. Fill in the blank with this word: ""It ___ Necessarily So""

51. Trading places: Abbr.

52. Presidential ___

53. Love ___

55. Ka ___ (Hawaii's South Cape)

PUZZLE 2

ACROSS

1. Charges anew

10. This longest river in Europe enters the Caspian by way of Astrakhan

15. Film box datum

16. Vast, old-style

17. Donizetti fan, e.g.

18. Fill in the blank with this word: "English philosopher George Henry ___"

19. When it's broken, that's good

20. University of Kentucky's ___ Arena

21. Sci. course

22. Verdi's "___ tu"

23. The supreme Supreme

25. Within walking distance

26. Outlawed blasts

29. Combined, in Compi

31. Fill in the blank with this word: "___ el Amarna, Egypt"

32. Swedish actress Persson

33. More swift

35. Vast

37. When fighting ends, in short

38. World's highest large lake

41. Wheelchair access

44. ___ Robbins, co-lyricist of the #1 "Rocky" theme song "Gonna Fly Now"

45. 1953 Oscar-nominated film based on a novel by Jack Schaefer

46. Handel's "___ in Egypt"

48. Vocal complaint

50. Fill in the blank with this word: "Andean peak ___ Cruces"

52. Put ___ good word for

53. What is the capital of this country - Switzerland

54. Part of A.A.U.W.: Abbr.

56. Singer ___ King Cole

57. Sci. of insects

59. Kind of park

61. Yellow parts

62. Budget director under Jimmy Carter

63. Fill in the blank with this word: "___ a fox"

64. Small piece of luggage

DOWN

1. The English translation for the french word: rouvrir

2. From one side only, in law

3. Silly tricks

4. Fill in the blank with this word: "Anderson's "High ___""

5. Tolstoy's "___ Fyodor Ivanovich"

6. Means

7. Mysterious art visible from the sky

8. Stays current

9. Ukraine, e.g., formerly: Abbr.

10. Speed: Abbr.

11. Jermaine ___, six-time N.B.A. All-Star

12. One hit by a tuba

13. Young tough

14. Popular brew from Holland

24. 1983 Indy 500 winner Tom

25. Los Angeles's ___-Sinai Medical Center

27. Fixed at an acute angle

28. Temp, often

30. Weapon for Iraqi insurgents: Abbr.

33. The English translation for the french word: insuffisant

34. Month after Nisan

36. Fill in the blank with this word: ""___ bin ein Berliner""

38. Product in a 1982 recall

39. In a sluggish way

40. Earwax

42. Fight night highlight

43. What some sinners do

44. Where brothers and sisters hang out

46. Puts one's foot down

47. Unscramble this word: rtatel

49. Minneapolis suburb

51. Harden

55. Soft palate

58. Works for an ed.

59. Fill in the blank with this word: "___ system"

60. Fill in the blank with this word: "___ tai"

PUZZLE 3

ACROSS

1. Shortchanged

7. Urban area in a Cheech Marin film

13. Turned, as topsoil

15. Mr. Blues player

16. Overdress, maybe

17. Tots

18. Tax

19. Kissy-faced

21. Paul Anka's '___ Beso'

22. Hydrocarbon suffixes

24. Does the sidestroke or butterfly

25. Suffix with origin

26. To-do list notations

28. Fill in the blank with this word: "___-Pitch"

29. Cornerstone abbr.

30. Stow, as cargo

32. Warning on an airplane wing

34. T-shirt size: Abbr.

35. Give ___ go

36. Turn red, perhaps

39. Terrifying

42. Head-___ (thorough)

43. Fill in the blank with this word: "Anderson's "High ___""

45. Prelate's title: Abbr.

47. Impose ___ on (outlaw)

48. Paul who won a Nobel in Physics

50. Oscar winner ___ Thompson

51. Fill in the blank with this word: ""Thanks a ___!""

52. Author better known as Saki

54. You can bob for apples because they're 25% this, which allows them to float

55. Song written by Queen Liliuokalani

57. Northeastern university where Carl Sagan taught

59. Vacation destinations

60. Small bone

61. St. Francis of ___

62. Kind of wrench

DOWN

1. Blowpipe emission

2. Some Arabs

3. Unscramble this word: roepctj

4. You might not be able to stand this

5. Old English letters

6. Regards

7. Parts of masks

8. Fill in the blank with this word: ""The House Without ___" (first Charlie Chan mystery)"

9. With 17-Down, a temporary urban home

10. Small, simple flute

11. Strong hand cleaner

12. Gather on the surface, chemically

14. Not stay alert

15. "You've gotta be kidding!"

20. World production of this is now 84 million barrels a day, up about 10 million barrels from 1997

23. Sheriff Deadeye creator

25. Semitic fertility goddess

27. Transition

29. To be, in Toledo

31. See 20-Across

33. Fill in the blank with this word: "Confit d'___ (potted goose)"

36. They may be suspended from art class

37. Wondering why

38. How often 55-/17-Across was married

39. Bygone Spanish dictator

40. LATE AUTHOR-HUMORIST

41. The Greatest Show on Earth' director

42. Soap opera actress Braun

44. Tulsa sch. named for an evangelist

46. Fill in the blank with this word: ""___ vile" (epithet for Falstaff)"

48. Hindu's loin cloth

49. Unscramble this word: ocrss

52. Winter frosts

53. Fill in the blank with this word: ""...___ they say""

56. Fill in the blank with this word: "___ polloi"

58. Robertson of CNN

PUZZLE 4

ACROSS

1. Quarters

5. Large-oared craft on a ship

13. Outfielder Mondesi

14. Smoke with straight sides

15. Fill in the blank with this word: "___ arch"

16. Is nosy

17. Rearranges the lettuce?

19. Corp. money managers

20. Alphabet sextet

21. Body of water in a volcanic crater, for one

23. The English translation for the french word: synode

25. Oust from office

29. Fill in the blank with this word: "ì___ boom bah!î"

32. With Altair and Vega, it forms the Summer Triangle

34. Money for Amer. allies

35. Zebulun's mother, in the Bible

37. Arrange into new lines

39. She, in Italy

40. Options at a gym

42. Wicked Game' singer Chris

44. Fill in the blank with this word: ""___ will be done""

45. Vowel sound in "puzzle"

47. Gradually quickening, in mus.

49. Fond ___, Wis.

51. Tricky fellows

55. Russian gold medalist ___ Kulik

58. Jungle gym's place

60. Items filling a star's mailbox

62. Prized game fish

63. Personal ad info

64. Zeno of ___

65. Some fortified wines

66. Twinkies or cookies, e.g.

DOWN

1. Erich who wrote "The Art of Loving"

2. Include as an extra

3. Ingredient in a Spanish omelet

4. Visits dreamland

5. Scented pouches

6. Jermaine ___, six-time N.B.A. All-Star

7. Your line of fate is quite deep, indicating success investing with tech stocks, like Adobe & Oracle, on this exchange

8. Ensured: Abbr.

9. Cradlesong

10. Two-time Swedish prime minister Palme

11. Fill in the blank with this word: "___ Snaps (dog treats)"

12. Russia's Itar-___ news agency

14. Oom-___ (polka rhythm)

16. Wage ___

18. Wallpaper meas.

22. Nickname of 1954 home run leader Ted

24. The ___ Love' (R.E.M. hit)

26. Yesteryear

27. They take the bait

28. Four hours on the job, perhaps

29. Stallone and others

30. Fill in the blank with this word: "Architect ___ Ming Pei"

31. Fill in the blank with this word: "___ palm"

33. U.K. carrier, once

36. Helm cry after "Ready about"

38. Flakes

41. Fill in the blank with this word: "Disco ___ (character on "The Simpsons")"

43. Fill in the blank with this word: "___-Ration (dog food)"

46. Woodstock, N.Y., county

48. They're no experts

50. Fill in the blank with this word: "___ part (role-plays)"

52. Actor Raf of "The Italian Job," 1969

53. Ohio natives

54. -

55. Fill in the blank with this word: ""___ not back in an hour...""

56. Pop singer ___ Del Rey

57. Having depth

59. Surgery sites, for short

61. Verdi's "___ tu"

PUZZLE 5

ACROSS

1. That's an order!

6. Ones graded E-8 in the Army

11. Vietnam War-era org.

14. There ___ stupid questions'

15. Burn slowly

16. Shamus

17. With 73-Across, be beaten by the rest of the field

19. Fill in the blank with this word: "Carolina ___"

20. Intensify

21. See 20-Across

23. It may finish second

24. Yield

27. Procedures: Abbr.

28. Strong cleaners

30. Turn red, maybe

32. Corp. money managers

33. Fill in the blank with this word: "___-frutti"

35. Drives off

37. Statistician's margin for error

39. Southern region of ancient Palestine

40. Fill in the blank with this word: ""Shall ___?" ("Want me to continue?")"

41. Nose: Prefix

42. Holy, to Horace

44. Warren ___, 1999 N.F.L. Defensive Player of the Year

48. Tie ___ (get smashed)

50. Tempur-Pedic competitor

52. Untilled tract

53. Part of the mailing address to Oral Roberts University

55. Inferences

57. This abbreviation .gov promises that the organization is "bringing safety to America's skies"

58. See 70-Down

61. Where the Paran

62. Soixante minutes

63. Not one's cup ___

64. Ham on ___

65. Volga region native

66. Big-time competition: Abbr.

DOWN

1. Labyrinthine

2. This word for a cantankerous personality is a variation of "ordinary"

3. Wee-hours trip

4. Fill in the blank with this word: "___ to one's neck"

5. Managed

6. Fill in the blank with this word: "___ Explorer (Web browser)"

7. T-shirt sizes, in short

8. Veer off track

9. TV/radio host John

10. Spanish counterparts of mlles.

11. Is sure not to miss

12. Old Fireflites and Firedomes

13. Straight run for skiers

18. Noble thing

22. They're loaded

25. Wine and dine

26. With 52-Across, what treasure hunters do

29. Some apartments

31. Ushered

34. Ulan-___, Russia

36. Water

37. It's the French word used for the body of an airplane minus the wings & tail

38. Teen-___

39. the first month of the year

41. Within walking distance, say

43. Try to see

45. Under the Lilacs' writer, 1878

46. Tube-nosed seabird

47. Succeed in appearing to be

49. Fill in the blank with this word: ""Stille ___" (Christmas carol)"

51. Wing

54. Olive genus

56. Tucson sch.

59. Veracruz Mrs.

60. Most miserable hour that ___ time saw': Lady Capulet

PUZZLE 6

ACROSS

1. Resinous juice used in perfumery

9. Large sea ducks

15. Stage actress Duse

16. Square, in 1950s slang, indicated visually by a two-hand gesture

17. A diva may throw one

18. Jazz virtuoso Garner

19. Meshed foundation in lace

20. Watchdog org.?

22. Enthusiastic snapper

28. You might hear it from someone who doesn't call

32. Maid of ___

33. Prepared to propose

34. Fill in the blank with this word: "Comic strip "___ &

Janis""

36. Dweeb

37. Stravinsky's "___ for Wind Instruments"

38. Singer Des'___

39. Port south of Osaka

40. Maelstrom

41. Mr. ___, radioactive enemy of Captain Marvel

42. Singer of the #1 country hit "Foolish Pride"

43. The English translation for the french word: imiter

45. If I Can't Have You' singer Elliman

47. Fastener for basement flooring, perhaps

49. Twilight

50. Get ___ (be rewarded at work)

56. It's a blessing

60. Finally turned (to)

62. Without flaw

63. Minus

64. The Supreme Court, e.g.

65. Uninteresting progress report

DOWN

1. Specialty oven

2. Sam Shepard's "___ of the Mind"

3. Porgy and ___'

4. Two tablets, maybe

5. Novelist Seton

6. Easy to prepare, say

7. Switzerland's Bay of ___

8. Island near Quemoy

9. Villain of Spider-Man

10. Handel's "___ in Egypt"

11. Wagner's "___ fliegende Hollander"

12. President Morales of Bolivia

13. Theologian's subj.

14. Fill in the blank with this word: "___ Digital Shorts (late-night comic bits)"

21. Ancient dynasty founder

23. Gives hope to

24. Win back, as trust

25. Pacino's "Sea of Love" co-star

26. Greek Catholic, e.g.

27. The Martian' garb

28. Household ___

29. Stylize anew, as a car seat

30. Either jamb of a doorway

31. TV's Cousin ___

33. "Born on the Fourth of July" hero Ron

35. Rock's ___ Speedwagon

39. Pou ___ (vantage point)

41. Synchronized

44. Painter ___ del Sarto

46. Pickle brand

48. Where Goodyear is headquartered

51. "Zuckerman Unbound" novelist

52. Greek gulf or city

53. Pack ___ (quit)

54. Hong Kong's Hang ___ Index

55. Fill in the blank with this word: "Competitive ___"

56. WorldCom competitor

57. Years on end

58. With 100-Across, Naples opera house Teatro di ___

59. Sport-___ (vehicle)

61. Weezer's music genre

PUZZLE 7

ACROSS

1. Superman's birth name

6. Nudges

10. The Sun and Mercury are in it: Abbr.

14. Spanish folk song

15. Fill in the blank with this word: ""Zip-___-Doo-Dah""

16. Part of Popeye's credo

17. Fill in the blank with this word: ""Victory ___" (1954 film)"

18. Fill in the blank with this word: "___ pit (rock concert area)"

19. Fill in the blank with this word: "___-shanter"

20. Celestial equator supergiant

22. Four hours on the job, perhaps

23. Usher's offering

24. Half of an old Latin aphorism

26. Writer Asimov

28. Take ___ sign

32. Latin lambs

34. Pianist born in Kurilovka, Poland

38. Unstable leptons

39. 'Je ne ___ pas franÁais'

40. Fill in the blank with this word: ""___ you not""

41. Party pooper who likes picnics in the rain?

43. Wallop

44. Vegetable fats

45. Sporty ride

47. Pinna's place

51. Fill in the blank with this word: "Debussy's "Air de ___""

54. the opposite of fat

57. Empathetic comment

59. Look at

60. Speck of dust

61. The English translation for the french word: sitar

62. Fill in the blank with this word: "___ Verde National Park"

63. Victor Nu

64. Trial figure

65. Tallinn native

66. Vex

67. Things gained and lost in football

DOWN

1. Most sacred building in Islam

2. Unscramble this word: laret

3. Under-the-wire

4. 'Waiting for the Robert ___'

5. Yellow Teletubby

6. He was Sonny to Marlon Brando's Vito

7. Two-time N.L. batting champ Lefty

8. Painting surface

9. To look, in Leipzig

10. O. Henry's 37-Across order?

11. Swimming great Diana

12. The Crimson Tide, familiarly

13. Treaty of Nanking port

21. Italian brandy

25. Fill in the blank with this word: ""Oh wad some power the giftie ___ us": Burns"

27. End of a cheer

29. "This isn't a good time"

30. Vaudeville bit

31. Presidential ___

32. Fill in the blank with this word: ""A one and ___...""

33. Highlander

35. Nickname for Dwight Gooden

36. Big shoe request

37. Open up a rip again

42. Y.A. Tittle's alma mater

46. Like Rapunzel

48. King of Tartary in "Turandot"

49. Worrisome food contamination

50. Honey badger

52. Fill in the blank with this word: ""Try ___ see""

53. Old Houston hockey team

54. Covered with many small figures, in heraldry

55. Strong cleaners

56. Y to the max?

58. Fill in the blank with this word: "___ fire (started burning something)"

PUZZLE 8

ACROSS

1. Most robust

9. Relative of a 29-Down

15. Fill in the blank with this word: ""___ pa? He feelin' better?""

16. Shredded

17. Some threats

18. The English translation for the french word: canapÈ

19. Six-foot vis-

20. 0,1

22. Fill in the blank with this word: "Basketball's ___ Elmore"

23. When some summer reruns are broadcast: Abbr.

24. Fill in the blank with this word: "Eye ___"

25. Yearbook signers: Abbr.

26. Student in 25-Down

30. Fill in the blank with this word: ""Scrubs" co-star ___ Braff"

32. Nickelodeon's Kenan and ___

33. Thistlelike plant

35. Southwest sidekick

37. Debate anew

40. Fill in the blank with this word: "___ line (elaborate barrier system)"

42. Wouldn't hurt ___

43. Very sorry

45. Painter's deg.

46. Theodore Roosevelt Natl. Pk. setting

48. Stiff hairs

51. Type of 35mm camera

52. Jacquerie

54. Went off

56. The English translation for the french word: vrombir

57. Makes teary

59. Cosmos star

60. Egg container

62. "Eureka!"

64. Returnee's question

65. Court encouragement

66. Worries

67. Petitions

DOWN

1. Terre ___, Ind.

2. Take cargo from

3. "Fiddler on the Roof" setting

4. Actor Dullea

5. Elementary suffix

6. Those: Sp.

7. Went after

8. a swing used by circus acrobats

9. Yadda, yadda, yadda'

10. Tobacco ___'

11. Fill in the blank with this word: ""Ol' Rockin' ___" (bin-mate of the 1957 album "Ford Favorites")"

12. Mukluk material

13. Id checker

14. Hirer's posting

21. It begins "In the Lord I take refuge"

27. Sein : German :: ___ : French

28. Unlikely steakhouse patron

29. Fill in the blank with this word: "___ minÈrale"

31. Fill in the blank with this word: "___ Tribunal (international court)"

34. Wrap in bright colors

36. Soft roe

37. Emphasizes with vehemence

38. Offensive smells

39. Chicken Little, for one

41. Military branches: Abbr.

44. Subject of the Brest-Litovsk treaty, 1918

47. Sweet to the ears

49. Fill in the blank with this word: ""___ Now" (1968 R & B album)"

50. Mama Cass ___

53. Monte ___ of Cooperstown

55. Small paving stones

57. The Crimson Tide, familiarly

58. The English translation for the french word: son

59. Unscramble this word: epos

61. Worrying sound to a balloonist

63. What Austrians speak: Abbr.

PUZZLE 9

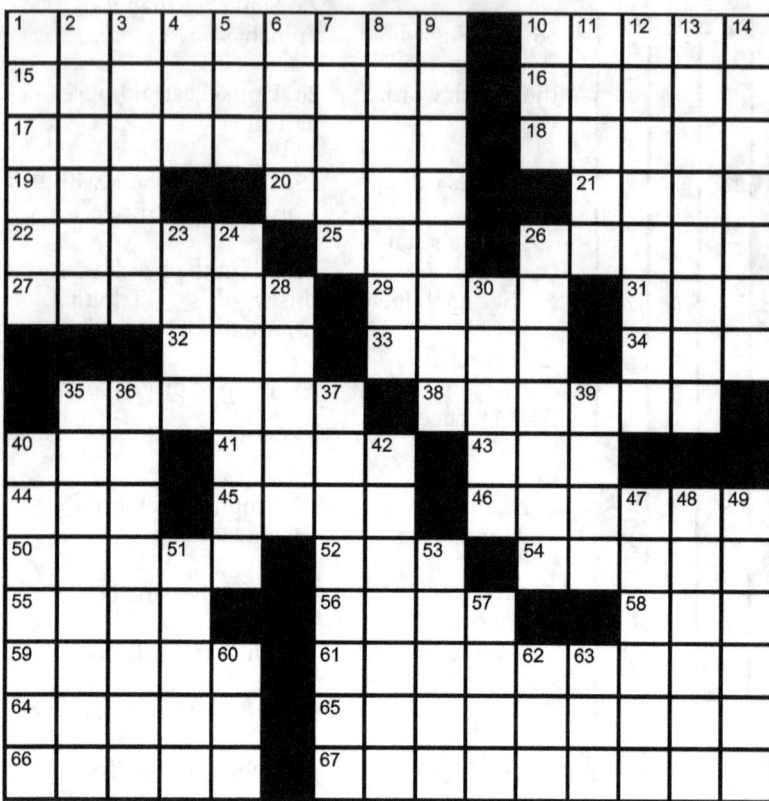

ACROSS

1. Station that's part of a TV network

10. Loafers holder

15. Rips into

16. Fill in the blank with this word: "___ flask (lab container)"

17. Another name for Tennessee

18. Fill in the blank with this word: "___ different tune"

19. Fill in the blank with this word: ""___ help a lot!""

20. Where the Gila joins the Colorado

21. Tiny time unit: Abbr.

22. With 62-Across, a possible title for this puzzle

25. Fill in the blank with this word: "Basketball's ___ Elmore"

26. The English translation for the french word: azur

27. Post-___

29. Haing S. ___ (Oscar winner for 'The Killing Fields')

31. Pay-___-view

32. Vitamin C source

33. Unscramble this word: nett

34. Taiping Rebellion general

35. Unscramble this word: plsaee

38. Fresh start, as of a movie series

40. Photography abbr.

41. West Coast sch.

43. Manute ___ of basketball

44. 1983 Indy winner Tom

45. Fill in the blank with this word: "___ Verde National Park"

46. The English translation for the french word: dorloter

50. Starfleet V.I.P.'s: Abbr.

52. QB Detmer and others

54. Ship's mooring aid [black]

55. Once I ___ secret love...'

56. Well-off

58. Three after B

59. TV's Gray and Moran

61. God-awful

64. Blusterous

65. Mission souvenirs

66. Oliver's Story' author Erich

67. Darjeeling, e.g.

DOWN

1. It's discrimination against & stereotyping of older people

2. Like a romantic dinner

3. Fill in the blank with this word: "___ bender (minor accident)"

4. Woe ___' (popular grammar book)

5. Star of "Youngblood," 1986

6. With double Fs you might play this word meaning full of uncertainty

7. Run ___ of (violate)

8. The English translation for the french word: tourmenter

9. Risk

10. Vietnam War-era org.

11. Pittsburgh-based food giant

12. Admits

13. Most ready to get started

14. a Mexican laborer who worked in the United States on farms and railroads in order to ease labor shortages during World War II

23. Useful Latin abbr.

24. Hymns of thanksgiving

26. Coffee table tome, perhaps

28. Rodents, jocularly

30. Year in Augustus Caesar's reign

35. "That Girl" girl

36. the sport of riding on a sled or sleigh

37. Sandwich filler

39. Veteran

40. Steers clear of

42. Flatter

47. Make extracts from by boiling

48. Nader's 2000 running mate

49. One of the Crusader states

51. Joe ___, ex-Royals third baseman known as the Joker

53. Young haddock

57. Fill in the blank with this word: "___ soit qui mal y pense"

60. Word part: Abbr.

62. Neckcloth

63. Lithium-___ battery

PUZZLE 10

ACROSS

1. They test reasoning skills: Abbr.

6. Windows frameworks

12. There are eight in a cup

14. Horace Walpole's "The Castle of ___"

16. Specters

18. Spells

19. Place runners?: Abbr.

20. Compresses, informally

22. Santa ___, Calif.

23. Fill in the blank with this word: ""___ Flux" (Charlize Theron film)"

25. Late actor Davis

26. Visibility reducer

27. Scott ___, 1997 N.L. Rookie of the Year

29. Robert Louis Stevenson's "___ Triplex"

30. The English translation for the french word: retomber

31. Patella

33. $&#@' and '%*&!'

34. Rodenticide name

35. Fill in the blank with this word: "___ bread"

36. Wild cards, in a certain game

39. Pilot's zone

43. Cold sorrel soup

44. Shingle abbr.

45. N.F.L. running back Barlow

46. Gulf of ___, body of water next to Viet Nam

47. Starts to raise, as a hem

49. Expires

50. Windows may have them, briefly

51. They cover the ears

53. Small island

54. Big Italian daily

56. Enjoyed a British tradition

58. Has as a customer

59. Least sweet

60. Colorist's vessel

61. You can have a lack of it in fashion, or use chemoreceptor cells to aid it at Chuck E. Cheese's

DOWN

1. D, for one

2. Walter Reed, e.g.

3. Novelist France of France

4. Cable TV giant

5. The English translation for the french word: thÈorie des ensembles

6. Fill in the blank with this word: ""___ Is Love" (1962 hit)"

7. Up ___ (trapped)

8. Women of Andaluc

9. Fill in the blank with this word: "___ Solo of "Star Wars""

10. Settles down for the night

11. Constrict, as a passage

13. Jerk

15. Some Siouans

17. Best-selling novelist about whom Gore Vidal said "She doesn't write, she types!"

21. Go on a vacation tour

24. Massachusetts city, birthplace of N. C. Wyeth

26. Mexican blankets, in M

28. SNL' network

30. Ukrainian city in W.W. I fighting

32. Long-running film role

33. Street sign abbr.

35. Card game for two

36. Classic work by Montaigne

37. Like some eaves in winter

38. Horrible

39. Fill in the blank with this word: "Alumni ___: Abbr."

40. Takes to the skies

41. Least likely to be pinned down

42. Put into office: Var.

44. Implant deeply

47. Fill in the blank with this word: "___-Bismol"

48. Trudge

51. North Carolina college

52. Work with mail

55. One in the charge of un instituteur

57. Fill in the blank with this word: "___ Motors"

PUZZLE 11

ACROSS

1. Yeah, sure!'

5. Revenge is ___ best served cold'

10. U.S.N. personnel

14. Sch. staffer

15. Tough, durable wood

16. Fill in the blank with this word: "___ quiet!"

17. Summer cooler

18. Yucky

19. Hubbub

20. Fill in the blank with this word: "Christie's "Death on the ___""

21. Strauss's "___ Heldenleben"

22. Bloomsbury group writer

23. Revolve

27. Physicist Ohm

30. Litigants

31. Negates

33. TV schedule abbr.

34. Fill in the blank with this word: "Colo. ___, Colo."

38. "I'm glad I came"

41. Welsh symbol

42. Title for a prince or princess: Abbr.

43. Error

44. Pierre's girlfriends

46. Praying figure

47. "That's enough out of you!"

52. Scythe handle

53. The English translation for the french word: ‡ma

54. Town NNE of Santa Fe

58. 1953 Oscar-nominated film based on a novel by Jack Schaefer

59. Word with Asia or Ursa

61. Vaulted space

62. South American monkey

63. Speedy sharks

64. Peer group?

65. Small barracuda

66. Fill in the blank with this word: ""Hello ___" (Todd Rundgren hit)"

67. Track units

DOWN

1. Pack ___ (quit)

2. Initials in a 1991 financial scandal

3. Jazzy Waters

4. Ring bearer

5. Writer Horatio

6. Singer Bobby and others

7. Improve one's golf game?

8. Barges

9. Hot spots

10. Winter warmer

11. Showy flowers

12. Old Greek coins

13. Lowly ones

22. Bleaches

24. Fill in the blank with this word: "___ fruit"

25. Biblical peak

26. Kaffiyeh wearers

27. Though spelled differently than her BFF, Ms. King, this is Oprah's middle name

28. Duck: Ger.

29. Beginning

32. Stupid jerk

34. It's usually tucked in

35. Rap's Salt-N-___

36. 1960 Updike novel

37. Library Card Sign-Up Mo.

39. ___ Church, country singer with the #1 hits 'Drink in My Hand' and 'Springsteen'

40. Wells's oppressed race

44. Fill in the blank with this word: "Carolina ___"

45. Violent Saharan wind

47. Attention getters

48. Square

49. Oncle's wife

50. Wiped out, slangily

51. The Norse language

55. What is the capital of this country - Samoa

56. Poet Mandelstam

57. Wrinkly-faced dogs

59. Year of Bush's swearing-in

60. Fill in the blank with this word: ""Am ___ risk?""

PUZZLE 12

ACROSS

1. Fill in the blank with this word: "Director Gus Van ___"

5. The red-spotted type of this salamander is one of the most common in the U.S.

9. Fill in the blank with this word: ""___, of golden daffodils": Wordsworth"

14. Paris's ___ d'Orsay

15. You never had ___ good!'

16. Tennis star Petrova

17. Thoroughly

20. Panama, e.g.

21. They lap France's coasts

22. Knives, forks and spoons

23. V-chips block it

24. Boxer Max

25. Inquire about a union contract?

32. The English translation for the french word: co-opter

33. Lawyers: Abbr.

34. Sportscaster Allen

35. Chaney Jr. and Sr.

36. Works for an ed.

37. Peer group?

38. Santa ___, Calif.

39. Onetime Chevy subcompact

41. Sell for

42. 1950 Ethel Merman song

46. Suffix with origin

47. The ___ 'e knows above a bit, the bullock's but a fool': Kipling

48. Yellow finch

50. Sacred ___

51. TV schedule abbr.

54. 1995 Annie Lennox hit

57. Winged

58. Marshal ___, Yugoslavian hero

59. 2nd qtr. starter

60. 1988 Peter Allen musical

61. The N.Y. Cosmos were in it

62. Yorkshire river

DOWN

1. Tiler's meas.

2. Fill in the blank with this word: ""___ Lee" (classic song)"

3. Sodium hydroxide, to chemists

4. Tiny ___

5. Some strong oxidizers

6. Forever, poetically

7. Circus reactions

8. Youngster

9. lues

10. John Wayne film set in Africa

11. The ___-Neisse Line

12. Your majesty'

13. Fill in the blank with this word: "Farmer's ___"

18. Fill in the blank with this word: "___-Hawley Tariff Act of 1930"

19. Work ___

23. Afterthought #3: Abbr.

24. Wrigley sticks?

25. Where many students click

26. TV's "20/20" creator Arledge

27. Chaplin and others

28. Trowel wielder

29. Peace

30. This insurance company features a gecko in its ads

31. Grant's first secretary of state ___ Washburne

36. Rabin's predecessor

37. Fill in the blank with this word: ""Able was ___...""

39. Not nude

40. MTV alternative

41. Best Musical of 1999

43. Wheel of Fortune et al.

44. Transforming Tonka toys

45. Trifle

48. Lemon ___

49. Webzine

50. Russian gold medalist ___ Kulik

51. Refill when you don't really need to

52. Stickers

53. Japon's place

54. Big inits. in paperback publishing

55. BBC rival

56. Ming of the Houston Rockets

PUZZLE 13

```
 1   2   3   4       5   6   7   8       9  10  11  12  13
14              ░   15          ░   16
17              ░   18          ░   19
20              21          ░   22              ░
░       ░   23          ░   24              ░       ░
25  26  27  28      ░   29              ░   30  31  32
33          ░   34          ░   35
36          ░   37          ░   38
39          ░   40          ░   41
42          43          ░   44          ░
░   45          ░   46          ░       ░
░   47  48              49          50  51  52  53
54          ░   55          ░   56
57          ░   58          ░   59
60          ░   61          ░   62
```

ACROSS

1. Common rhyme scheme

5. Fill in the blank with this word: ""I know not why I ___ sad": Shak."

9. Rounded end

14. The muse of history

15. Fill in the blank with this word: "___ Inn"

16. Yeshiva leader

17. Fill in the blank with this word: "Dog : paw :: horse : ___"

18. To pursue for food or sport

19. Fill in the blank with this word: "___ the side of caution"

20. One magazine's view?

23. Wayne LaPierre's org

24. Wild ___

25. Partner

29. Magnum ___

30. Scott's "___ Roy"

33. Where Prince Philip was born

34. Actress Scala and others

35. Young socialites

36. Tent used by a Latin musician?

39. Those, to Tom

40. Van ___, 'Lane in Autumn' painter

41. Witherspoon of "Vanity Fair"

42. TV pooch

43. When doubled, popular 1980s-'90s British sitcom

44. Swelled

45. Vivacity

46. Range part: Abbr.

47. "I didn't understand a thing you said"

54. This seaside structure is abbreviated whf.

55. Fill in the blank with this word: ""___ show you!""

56. Four hours on the job, perhaps

57. Belgian composer Guillaume

58. Unscramble this word: seem

59. Unusual shoe spec

60. "Look west," to a drill sergeant

61. Smartphone introduced in 2002

62. Tell ___ story

DOWN

1. Worms cries

2. Golden ___ (century plant)

3. High school subj.

4. Theatrical hit, in slang

5. The English translation for the french word: correspondre

6. Spanish actress Carmen ___

7. Lip-___

8. Muscle manipulator

9. QB Favre and others

10. Persian fairies

11. Zaragoza's river

12. The English translation for the french word: ÈbËne

13. Pitcher Robb ___

21. Starts to raise, as a hem

22. VCR button

25. Disperse

26. Sang-froid

27. "What the Butler Saw" playwright

28. Saturn vehicles?

29. Rock's ___ Boingo

30. Replace a wooden pin

31. Wide-bodied

32. Pet shop bagful

34. a specialist in geology

35. Unscramble this word: dere

37. Wrinkly fruit

38. Ventriloquist's prop

43. pawky (similar term)

44. Sailor's protector

45. Uncovers

46. The English translation for the french word: mÍlÈe

47. Fill in the blank with this word: ""___ Say," 1939 #1 Artie Shaw hit"

48. Cod's cousin

49. Critic, at times

50. TV's "___-Team"

51. Purcell's "___ and Welcome Songs"

52. San ___, Calif.

53. Look at

54. Corduroy ridge

PUZZLE 14

ACROSS

1. White ___ (termites)

5. Fill in the blank with this word: "___ even keel"

9. Fill in the blank with this word: "___ grabs"

14. U.K. carrier, once

15. Fill in the blank with this word: "___ de boeuf"

16. Fill in the blank with this word: "Elaine ___ ("Taxi" role)"

17. 2000 Supreme Court case hinging on the 14th Amendment

19. Scottish slopes

20. Tired? Pale? You may have this condition meaning "want of blood" & usually caused by an iron deficiency

21. Wobbly

23. Tony-nominated choreographer White

25. Lumberjack's tool

26. How jewelers get absolution?

33. Jennifer Lopez album 'J to ___ L-O!'

34. Finish this popular saying: "The bottom line is the bottom_____."

35. Troy, in poetry

36. Straight: Prefix

38. Sniggled

41. Fed. lending agency

42. successful (similar term)

44. Keto-___ tautomerism (organic chemistry topic)

46. They often accompany logos: Abbr.

47. Lament, part 2

51. The Greek Letter Equivalent : R

52. Every 12 mos

53. showing malicious ill will and a desire to hurt

58. Pops

62. Unlawful firing?

63. Nips

65. Fill in the blank with this word: "___ the Hedgehog (video game)"

66. Viennese-born composer ___ von Reznicek

67. One-named German singer who was a one-hit wonder

68. Outsider, in Hawaii

69. Fill in the blank with this word: ""Get ___!""

70. Some valuable 1920s-'40s baseball cards

DOWN

1. Title for some bishops

2. Proper ___

3. Fill in the blank with this word: ""Don't ___ me, bro!""

4. Nimrod

5. Curtain fabric: Var.

6. Emergency call

7. The Bell of ___' (Longfellow poem)

8. Dolce far ___

9. Puccini aria

10. Fill in the blank with this word: "___ zoologique (French zoo)"

11. Fill in the blank with this word: "___ Bl"

12. The ___-Neisse Line

13. Mateus ___

18. Record albums, to collectors

22. RR stop

24. Nary ___

26. Missing Links : Hammer ___ rug

27. Kind of yoga

28. Pequod's co-owner

29. Subj. with unknowns

30. Fill in the blank with this word: "Bottom of the ___"

31. Midsection, informally

32. Shirt sizes

33. Singer Tennille

37. Park activity

39. Fill in the blank with this word: ""Just you wait, ___ 'iggins...""

40. a push button at an outer door that gives a ringing or buzzing signal when pushed

43. 'Wait Until Dark' director Young

45. Whoppers

48. TV band above channel 13, in brief

49. Unhappy one

50. Rostand protagonist ___ de Bergerac

53. Window part

54. Swift Malay boat

55. This __ laughing matter!'

56. Work like Tillie?

57. Weak, as an excuse

59. Fixed at an acute angle

60. Windshield option

61. Hydros : England :: ___ : U.S.

64. Fill in the blank with this word: "___ Motors"

PUZZLE 15

ACROSS

1. Wrists, anatomically
5. Digital video file format
9. Unable to do well
14. Some CBS forensic spinoffs
15. Words of discovery
16. Grads
17. Chocolate source
19. Winter sight at Tahoe
20. Wunderkind
21. Poker great Ungar and others
23. "Amen!": 3 wds.
25. Vacationer's help
30. Kind of chair
32. Ancient Semite
33. Make ___ of things
36. Station wagon, in England
38. Fill in the blank with this word: "___ Tamid (synagogue lamp)"
39. Fill in the blank with this word: ""Ars gratia ___""
40. Har-___ (tennis surface)
41. He was Sonny to Marlon Brando's Vito
44. Give ___ (care)
46. How some dares are done
47. Renaissance instrument
49. W.W. II carrier praised by Churchill for its ability to "sting twice"
51. Ear ornaments
54. Fill in the blank with this word: "Basket-of-___ (yellow perennial)"
56. Prefix with bacteria
58. Woman's shoulder wrap
62. One mile, for Denver
64. Slippery as ___
65. Twining stem
66. Wide-lapel jackets
67. Samuel Gompers's org., informally
68. Zip
69. Neural network

DOWN

1. Soyuz rocket letters
2. Japanese beer brand
3. They're chosen for your sake
4. Manners
5. Wrestler
6. Sentence part: Abbr.
7. Go-aheads
8. 'Whatsoever ___ upon the belly ...': Leviticus 11:42
9. Sting, e.g.
10. Litmus bluer: Abbr.
11. What a weaver may be guilty of, briefly
12. French soul
13. Original Dungeons & Dragons co.
18. Chaplin and others
22. Fill in the blank with this word: ""___ directed""
24. Brutus's burdens
26. Taoism founder Lao-___
27. Anatomical roofs
28. Prepare to fight
29. What many incumbents do
31. To be, in Toledo
33. Winter pear
34. Wherewithal
35. Fill in the blank with this word: ""Aunt ___ Cope Book""
37. Like some harrows
39. Fill in the blank with this word: ""Don't ___ surprised""
42. Longtime essayist for The New Yorker
43. Though Utah's not on the ocean, this is its state bird
44. UnitedHealth rival
45. Antwerp artisan
48. Mark your card!
50. Newbie, of sorts
52. Notwithstanding
53. Radical
55. Mae West role
57. Beginning
58. This abbreviation .gov promises that the organization is "bringing safety to America's skies"
59. Like Brahms's Symphony No. 3
60. Wharton grad's aspiration, maybe
61. Publishing mogul, for short
63. Spanish queen until 1931

PUZZLE 16

ACROSS

1. His and ___

5. Red ___

8. used (similar term)

13. Talk of the Gaelic

14. Masked critter

16. Whale finder

17. Fill in the blank with this word: "___ the way"

18. The Flintstones' pet

19. Zeal

20. It's classified

23. Light ___

24. Yo te ___'

25. Mandela's land: Abbr.

28. Like a test with a properly corrected score?

33. Trickiness

36. Teacher's deg.

37. Fill in the blank with this word: "___ oil (perfumery ingredient)"

38. On the right

41. Cafeteria worker's wear

42. Shaped, as wood

43. Self: Prefix

44. Winding road shape

45. Chekhov play, with "The"

49. Troop grp.

50. Taiping Rebellion general

51. Fill in the blank with this word: "___ no"

55. Extra-base hit, probably?

60. Fill in the blank with this word: ""___ Irish Rose""

62. Step ___!'

63. Have ___ in one's bonnet

64. Fill in the blank with this word: "Caput ___ syndrome (arm problem)"

65. No-good

66. Cubism pioneer Juan

67. Red Sox Hall-of-Famer Bobby

68. Wenders who directed 'Buena Vista Social Club'

69. Twilights, poetically

DOWN

1. Up on, as the jive

2. Fill in the blank with this word: "___ Good Feelings"

3. Sent regrets, say

4. Vacillate

5. Fill in the blank with this word: ""___ Live," 1992 multiplatinum album"

6. Pinot ___ (wine)

7. Needle holder

8. Golf's ___ Aoki

9. L'...toile du ___, Minnesota's motto

10. Too pink, say

11. Fill in the blank with this word: "___ TomÈ"

12. To ___ is human ...'

15. You've got the wrong guy!'

21. Fill in the blank with this word: "Electric ___"

22. Related word

26. Monica ___, two-time U.S. Open champ

27. Miners' entries

29. Results in

30. Under 100 mg per deciliter of this is considered optimal

31. Retailer with stylized mountaintops in its logo

32. Ticket abbr.

33. Witch of ___

34. Rabin and Remini

35. Coin collector's classification

39. Fill in the blank with this word: ""Thanks for ___ Memory""

40. Singer Des'___

41. Pizza ___

43. Currently

46. Supports

47. Ham on ___

48. E.P.A. concern

52. This cavalry weapon was inspired by the Turkish scimitar

53. Vegetable oil component

54. Like non-oyster months

56. Shakespearean king

57. Fill in the blank with this word: "Battle of the ___, opened on 10/16/1914"

58. South American monkey

59. Flightless bird: Var.

60. I.R.S. employee: Abbr.

61. Slo-___ fuse

PUZZLE 17

ACROSS

1. That you should feed a cold and starve a fever, and others

5. Love, honor and ___

9. Ohio political dynasty

14. On ___ streak (winning)

15. To laugh, to Lafayette

16. Less refined

17. Tiny battery

18. No Clue

19. TV's Gray and Moran

20. South-of-the-border border town portmanteau

23. The English translation for the french word: marche

24. Fill in the blank with this word: ""Wait a ___!""

25. Winds up

28. Small songbirds

30. White Sulphur ___, W. Va.: Abbr.

33. Swiss dish of grated and fried potatoes

34. Chinese dynasty up to A.D. 1125

35. Sound of an air kiss

36. "Listen!"

39. One ___ at a time

40. Officially listed: Abbr.

41. Not yet solidified

42. Old what's-___-name

43. Jon of TV's 'Homicide'

44. The last arrow fired in an archery contest, today it means the final result

45. "The Conning Tower" columnist's inits.

46. Without restraint

47. Basic

54. Fill in the blank with this word: "___ artery"

55. Fill in the blank with this word: "___-ran"

56. Opera conductor Daniel ___

57. Fill in the blank with this word: "A votre ___"

58. You have the right to do this regarding arms, but your arms will be this without sleeves

59. Fill in the blank with this word: "___ avis"

60. 1988 Peter Allen musical

61. Fill in the blank with this word: "___ Mawr, Pa."

62. Western Electric founder ___ Barton

DOWN

1. Thank-you-___

2. Those, to Robert Burns

3. The English translation for the french word: intox

4. Part of a flight

5. They may have many stops

6. Microsoft chief, to some

7. Riley's "___ Went Mad"

8. Quaint affirmative

9. Some natural history museum displays, for short

10. Gold-related

11. Savings acct. protector

12. Fill in the blank with this word: "___'clock"

13. Yearbook signers: Abbr.

21. Make aware

22. Thwart in court

25. Wall Street debacle

26. Soprano Lehmann

27. Mixer maker

28. New York county whose seat is Owego

29. Fill in the blank with this word: ""___ my Annabel Lee": Poe"

30. Swagger

31. Madrid's ___ del Prado

32. Scarlett's love

34. Wasn't straight

35. No place for skirts

37. Finish this popular saying: "Fools rush in where angels fear to ___."

38. Kind of payment

43. Zoned (out)

44. unhatched (similar term)

45. Guinness Book listings

46. Start of an opinion

47. of bluish-black or grey-blue

48. Unscramble this word: rnig

49. Common rhyme scheme

50. What an A is not

51. Fill in the blank with this word: "Air___, discount carrier"

52. Fill in the blank with this word: "___ Aarnio, innovative furniture designer"

53. Transfer and messenger materials

54. Map abbr.

PUZZLE 18

ACROSS

1. Make ___ at

6. The English translation for the french word: suie

10. Unscramble this word: eyrv

14. Spasm

15. Amazes

16. Would ___?'

17. Tub filler

18. Sound stressed, maybe

19. Teutonic turndown

20. IRIS

23. Source of some rings

24. Riley's "___ Went Mad"

25. I ___ Rock'

28. Forty-___

31. Snow White' dwarf

35. ___ Institute (astronomers' org.)

37. Tampico track transport

39. Port of old Rome

40. Health V.I.P.'s

43. Yoga posture

44. Trickle out

45. Fill in the blank with this word: "___-majestÈ"

46. Testify under oath

48. Old English letters

50. Ways around: Abbr.

51. Like some muscles

53. Royal ___ (Detroit suburb)

55. *"Star Wars" actress who's a Harvard grad

62. William ___, Alaska's first 72-Down

63. Would-be J.D.'s hurdle

64. Where to get a date?

66. Rare book dealer's abbr.

67. Tot's injury

68. B. & O. stop: Abbr.

69. Whitewash

70. Actress Rowlands

71. Punctilious type, slangily

DOWN

1. Up

2. Fill in the blank with this word: ""___ Lap" (1983 film)"

3. Greek gulf or city

4. Old Buckeye State service station name

5. Like many linings

6. Jerk

7. It hurts!'

8. Sufficiently old

9. Met men

10. Fill in the blank with this word: ""The Little Engine That Could" as read by actor ___"

11. 'Waiting for the Robert ___'

12. New York's Jacob ___ Park

13. Urges

21. "Otello" librettist

22. French actor Alain

25. Syrian president Bashar al-___

26. Scene of W.W. I fighting

27. Set ___

29. Sea eagles

30. Watch again

32. ___ humains (people, in Paris)

33. First dynasty of Polish rulers

34. Some locks

36. Fill in the blank with this word: ""No nation is permitted to live in ___ with impunity": Jefferson"

38. See 1-Across

41. What las novelas are written in

42. Ancient Spartan magistrate

47. Postscript

49. Zen illumination

52. Fill in the blank with this word: ""___ Have No Bananas""

54. The first one opened in Detroit in 1962

55. Fill in the blank with this word: "Actor ___ Patrick Harris"

56. Latin lambs

57. Loc. of some devils

58. When the brain has incoming signals of this blocked in the spinal cord, the result is analgesia

59. Fill in the blank with this word: "Cup ___ (hot drink, informally)"

60. Fill in the blank with this word: "___ prof."

61. TV's Nick at ___

65. The English translation for the french word: DAS

PUZZLE 19

ACROSS

1. Unscramble this word: lsle

5. Now see ___!'

9. See 26-Across

13. I never ___ man ...'

14. Hebrew letter

15. Unscramble this word: htrdi

16. Sport involving a chute

18. Summer fabric

19. Tough customer

20. Winetaster's asset

21. Fill in the blank with this word: ""___ was saying Ö""

22. Gets in shape

23. Doctor's assurance

28. Winningest southpaw in major-league history

29. Switch suffix

30. Org. that produces the Congressional Record

33. Zeppo, for one

34. UPS rival

36. Naturalist Fossey

37. On the ___

38. Wife of Siva

39. La ___ vita

40. Treat

43. One new to a line

46. Fill in the blank with this word: ""___ note to follow ...""

47. Steamy, say

48. Cause for a reprimand from a teacher

53. Ticked off

54. Open the windows in

55. Pertaining to hair

56. Three-time speed skating gold medalist Karin

57. Keglers' places

58. Requests for developers: Abbr.

59. Made a tax valuation: Abbr.

60. Wearers of four stars: Abbr.

DOWN

1. With 52-Across, what angels pray for

2. Reggae's ___-Mouse

3. Long dist.

4. "This Gun for Hire" star

5. Lake ___ City, Ariz.

6. Fill in the blank with this word: ""L'___ d'Amore" (Donizetti opera)"

7. Oscar-winning French film director ___ Cl

8. Trio after D

9. Swiss quarters?

10. Succotash ingredients

11. Verdi baritone aria

12. Take ___ breath

15. Romantic plotter in "The Taming of the Shrew"

17. Raider Carl

20. Part of many addresses

22. Weight not charged for

23. Circus reactions

24. Flame Queen ___ (famous gemstone)

25. Fill in the blank with this word: "___ a soul"

26. Fill in the blank with this word: "1971 sci-fi film "___ 1138""

27. Kind of a drag

30. Fill in the blank with this word: "___-edged"

31. White-spotted rodent

32. Perfect report card spoiler

34. Stunning blow

35. Sommer in the cinema

36. John ___

38. More benevolent

39. Fill in the blank with this word: "Abu ___"

40. Vatican artworks

41. Sets (down)

42. Made some lace

43. Rubbish

44. Fill in the blank with this word: "___ to go"

45. The Beatles' "Any Time ___"

48. 1936 Olympics hero

49. "The Producers" role

50. In South Africa 100 cents equals R1, R standing for this

51. Watch part

52. Fill in the blank with this word: "___ Trueheart of "Dick Tracy""

54. Miles of film

PUZZLE 20

ACROSS

1. New York University's ___ School of the Arts
6. It's the 4-letter term for the thin sheets of dried seaweed in which sushi is wrapped
10. Wow
14. Rodgers & Hammerstein's "All I Owe ___"
15. Filmmakers: Joel & Ethan
16. Subject, in Spain
17. Prepare to fight
18. Place to make a scene?
20. Provisions in Hell?
22. Fill in the blank with this word: ""___ Marner""
23. Perfect start?
24. Rock-___, classic jukebox
27. Peter ___ Tchaikovsky
30. How often 55-/17-Across was married
32. 1940's-50's All-Star Johnny
35. Take-out order?
37. -
38. Fill in the blank with this word: "Entr'___"
39. Fill in the blank with this word: "___ Good Feelings"
41. 1960 Updike novel
42. Prefix meaning "likeness"
44. Sicilian resort city
45. She, in Italy
46. Bend out of shape
48. Guiding light
50. Fill in the blank with this word: "___ Monroe, "Green Acres" role"
51. Winning Super Bowl XXXVII gridder
53. Wake Up Little ___' (#1 Everly Brothers hit)
56. Pop singer known as "The Delta Lady"
60. Beatles song with a complaining title
63. Red as ___
64. Fill in the blank with this word: "___ burn"
65. Fill in the blank with this word: "___ me tangere"
66. Arriving at the tail end / Survive
67. Merle Haggard's "___ From Muskogee"
68. Onetime big inits. in car financing
69. Seed's exterior

DOWN

1. Undecided, you might say
2. Actress Skye and others
3. Wise guy
4. 23-Across hit
5. It's bound to be used in a service
6. Wildcats' org.
7. Fill in the blank with this word: "Astronomy's ___ cloud"
8. Fix, as a pool cue
9. Step on it
10. Poker great Ungar and others
11. Fill in the blank with this word: "___ Zeppelin"
12. Fill in the blank with this word: ""___ Blue?" (1929 #1 hit)"
13. Ming of the Houston Rockets
19. Woodworking fasteners
21. Where to sign a credit card, e.g.
24. Biblical dry measures
25. The Hare
26. Yoga posture
28. Smear with wax, old-style
29. Midwest and Plains states, e.g.
31. Impossible to change
32. Fill in the blank with this word: ""You ___ mouthful!""
33. Gradually quickening, in mus.
34. Words with hit or take
36. Years on end
40. String bean's opposite
43. What a satellite may be in
47. Silencing
49. One way to serve café
52. Two strikes?
54. Fill in the blank with this word: ""Gibraltar may be strong, but ___ are impregnable": Emerson"
55. Get rid of
56. Investment firm T. ___ Price
57. That, in France
58. Of lyric poetry
59. Jazz singer ___ James
60. Uniform: Prefix
61. Fill in the blank with this word: "___ Day, Jan. celebration"
62. Fill in the blank with this word: ""___ lied!""

PUZZLE 21

ACROSS

1. Fill in the blank with this word: ""Where there's ___ ...""

6. Glove

10. Wharton offerings: Abbr.

14. Tree with pods

15. Queen Anne's ___

16. Would ___?'

17. Hodgepodge

19. Tale of adventure

20. Bony fish

21. Trinidad's Pitch Lake is a natural lake of this black substance used to surface roads

23. Keglers' places

25. Towel ends?

26. Mil. V.I.P.

30. Fill in the blank with this word: "___-andrew (buffoon)"

33. Fill in the blank with this word: "Cutty ___ whisky"

34. Worth

35. Fill in the blank with this word: "___ roll"

38. You can't enjoy this if you've lost your marbles

42. Stock units: Abbr.

43. Mandela's presidential successor

44. LaSalle of "ER"

45. Snap

46. Curtain fabrics

48. Fill in the blank with this word: ""___ honor""

51. Word repeated in "Now ___ away! ___ away! ___ away ...!"

53. Option not available in a convertible

56. Where to find the diving board

61. The ___ Love' (R.E.M. hit)

62. Container for an iron or wedge?

64. Metallica drummer Ulrich

65. Anklebones

66. Joe ___, ex-Royals third baseman known as the Joker

67. Tallinn native

68. Movie whose hero is named Z-4195

69. Abbr. after Ted Kennedy's name

DOWN

1. Fill in the blank with this word: "___ no."

2. Tech expert, as it were

3. Suffix with poet

4. While away the time

5. The Ramblers of the N.C.A.A.

6. Instrument with fingerholes

7. Fill in the blank with this word: ""Bali ___""

8. With a 2007 women's water polo title, this California school became the first to win 100 NCAA titles

9. Vets' concerns

10. Powerful

11. Vocal complaint

12. Window alternative

13. Small paving stones

18. Big elephant features

22. Life: A User's Manual' author Georges

24. Kissed noisily

26. Pouches

27. Salt Lake City hoopsters

28. Michael Moore's "Downsize ___!"

29. Fill in the blank with this word: "___ Explorer (Web browser)"

31. Textbook market shorthand

32. Poe's "The Murders in the ___ Morgue"

34. Political second banana

35. Zilch

36. Crepe paper feature

37. Concert prop

39. See 81-Across

40. Org. for mom-and-pop stores

41. Fill in the blank with this word: ""The Longest Day" director ___ Annakin"

45. The English translation for the french word: tape-‡-líúil

46. Tiny time unit: Abbr.

47. "It has come to my attention ..."

48. Sicilia e Sardegna

49. Yellowfin and bluefin

50. Hagar the Horrible's dog

52. Madison Ave. trade

54. Two times tetra-

55. Cr√*me caramel : France :: ___ : Spain

57. Tot toter

58. Fill in the blank with this word: ""Empedocles on ___" (Matthew Arnold poem)"

59. Some yeses

60. Strands of biology

63. Last: Abbr.

PUZZLE 22

53. Perturb

54. Tip

55. Holmes and Moriarty, e.g.

56. Spring times

57. Give an interpretation or explanation to

58. Alibi

DOWN

An inflammation of the mucous membrane lining the nose

2. English artist noted for a series of engravings

3. Dragonflies and damselflies

4. Peewee

5. Confined, with "up"

6. A system of belief based on mystical insight into the nature of God and the soul

7. Engine parts

8. Simultaneously

9. A woodworker who joins pieces of wood with a splice

10. Of something totally lacking in saturation and therefore having no hue

11. Come to

12. "The Canterbury Tales" pilgrim

13. Having the shape of a sphere or ball

14. Amount of hair

21. Home to nearly 70% of all people

24. Primordial matter

28. Create laminate by bonding sheets of material with a bonding material

30. Equal

32. A refund of some fraction of the amount paid

33. A person who operates a farm

34. When your clothes wear out?

35. Sent signals to

37. Token

38. The rider of a horse used in eventing

39. The baby's room

40. Arctic ___

42. Marsh bird

44. Equals

49. "___ on Down the Road"

51. Cousins of the ostrich

52. Atomizer output

ACROSS

1. Teeth

6. Garbage

11. Made of, containing or resembling wood

13. Track orb

15. W.W. I battle locale

16. Made of, containing or resembling wood

17. A grey or greenish-blue mineral consisting of aluminum silicate.

18. Fighting

19. "The Snowy Day" author ___ Jack Keats

20. Flashed signs

22. Frosts, as a cake

23. Dapper

25. 100 cents

26. Bungle, with "up"

27. "I ___ return"

29. A murderer who slashes the victims with a knife

31. Otalgia

33. A metric†unit of weight equal†to one thousandth of a kilogram.

36. Aden's land

40. Face-to-face exam

41. Wading birds of warm regions having long slender down-curved bills...

43. Increase

45. Hamster's home

46. Indian bread

47. ___ mortals

48. Beseech

50. Titanic

PUZZLE 23

ACROSS

1. Rodrigo ___ de Vivar (El Cid)

5. Res ___ loquitur (legal phrase)

9. Relatively cool sun

14. Fill in the blank with this word: "Competitive ___"

15. Fill in the blank with this word: "___ guy (one who gets things done)"

16. Paris's ___-de-Medecine

17. Like the U.S. legislature

19. Sure competitor

20. Starbucks offering

21. Habituates

22. Transfer ___

23. Tracy

24. Pal

28. Like Brahms's Symphony No. 3

29. Saudi monarch

33. That is __...'

34. Perfect report card spoiler

35. Fill in the blank with this word: "China's Chairman ___"

36. Rebel

40. Fifth-century date

41. Golf's ___ Aoki

42. ___ shame (abash)

43. Book in which the destruction of Samaria is foreseen

45. Confounded

46. The English translation for the french word: milieu

47. Former Cub ___ Sandberg

49. Suffix with Mozart

50. Underwater trap

53. High card up one's sleeve

58. Fill in the blank with this word: "Arthur Miller play "___ From the Bridge""

59. Coupon for the needy

60. First name in mystery writing

61. Words often before a colon

62. Switch suffix

63. Yooks' and Zooks' creator

64. Fill in the blank with this word: ""The first ___, the angel did say ...""

65. Twilights, poetically

DOWN

1. Young socialites

2. Start to -matic

3. Part of 16-Across: Abbr.

4. Passion

5. Fill in the blank with this word: ""___ kick from champagne...""

6. Fill in the blank with this word: "Della ___ (St. Peter's architect)"

7. Union member

8. Yahoo! competitor

9. The Matrix' star Reeves

10. Back of the neck

11. Unit of pressure

12. Sam Shepard's "___ of the Mind"

13. Some wines

18. Thomas Becket, e.g

21. Like some private dets.

23. You're __ talk!'

24. #2 at the 1994 U.S. Open

25. Unilever skin cream brand

26. Late actor Davis

27. Writer Santha

Rama ___

28. They believed the world was created by Viracocha

30. The Brothers ___ (violinmakers)

31. Writer Bret

32. Fill in the blank with this word: ""___ want to dance?""

34. Spanish direction

37. Smiling

38. Thinks out loud

39. Void, in Vichy

44. Partner of jeweler Van Cleef

46. Comic Howie

48. Mournful wails

49. "More!"

50. St. Louis's historic ___ Bridge

51. Fill in the blank with this word: ""A Letter for ___" (Hume Cronyn film)"

52. In ___ of (replacing)

53. Prefix with sphere

54. Way from Syracuse, N.Y., to Harrisburg, Pa.

55. Fill in the blank with this word: "Europe's Gorge of the ___"

56. "Time's a-wastin'!"

57. Works of Homer

59. Underwater steerer

PUZZLE 24

[crossword grid]

ACROSS

1. Sea eagles

5. Increases greatly, as prices

9. Welcome sight for a castaway

14. Fill in the blank with this word: "___ de soie (silk cloth)"

15. Fill in the blank with this word: ""Die Frau ___ Schatten" (Strauss opera)"

16. Slightly above average

17. Kind of sale

19. Trifles: Fr.

20. White ___ ghost

21. 1977 double-platinum Steely Dan album

22. Picks up

23. Mandela's land: Abbr.

24. Mystery writer Gardner et al.

26. Swell

29. Redundancies, like 20- and 50-Across and 5- and 29-Down

33. Have ___ for (desire)

34. Fatty

35. Withdraw gradually

36. Food flavoring brand

37. Wielder of the sword Tizona

38. Vapory beginning

39. Fill in the blank with this word: "___-Honey (candy name)"

40. Opera singer

41. "That was close!"

42. Water coolers

44. The English translation for the french word: piste de ski

45. Fill in the blank with this word: ""___ fast!""

46. Union ___: Abbr.

47. Gems, precious metals, etc., in Spain

50. Writer Santha Rama ___

51. School media depts.

54. The English translation for the french word: trait

55. See 20-Across

58. Ukrainian port, to natives

Simon ___

59. Answer to the riddle "Dressed in summer, naked in winter"

60. Tennis's Mandlikova

61. Lulls

62. Supplementary: Abbr.

63. She-bears, south of the border

DOWN

1. Biblical dry measure: Var.

2. Actor Stephen et al.

3. "The Lion King" lion

4. "-er" or "-ing," e.g.: Abbr.

5. Sara portrayer on 'CSI'

6. Some Muslims

7. The Tar Heels: Abbr.

8. The Carolinas' ___ Dee River

9. Window insert

10. Actress Zadora visited Samoa's capital?

11. The Sopranos' actor Robert

12. Sally ___ (teacake)

13. Sound of a leak

18. Yesteryear

22. London insurance giant

23. EIOSN

24. Villain of Spider-Man

25. Knots again

26. Fill in the blank with this word: ""___ Meets Godzilla" (classic 1969 cartoon)"

27. With 39-Across, 21-/28-Across, for one

28. Spanish direction

29. Beats

30. Actor Green and others

31. "The Old Religion" novelist

32. Senator from Maine

34. Fill in the blank with this word: ""Never ___ tell thy love": Blake"

43. Main lines

44. Spittoon sound

46. Savanna region stretching from Senegal to Chad

47. 'Vette option

48. Wagner's earth goddess

49. Writer's supply: Abbr.

50. Snow White's sister

51. Fill in the blank with this word: ""___, 'tis true I have gone here and there": Shak."

52. Fill in the blank with this word: "___ cava"

53. Women of Andaluc

55. This abbreviation .gov promises that the organization is "bringing safety to America's skies"

56. La Guardia : LGA :: O'Hare : ___

57. Notwithstanding that, informally

PUZZLE 25

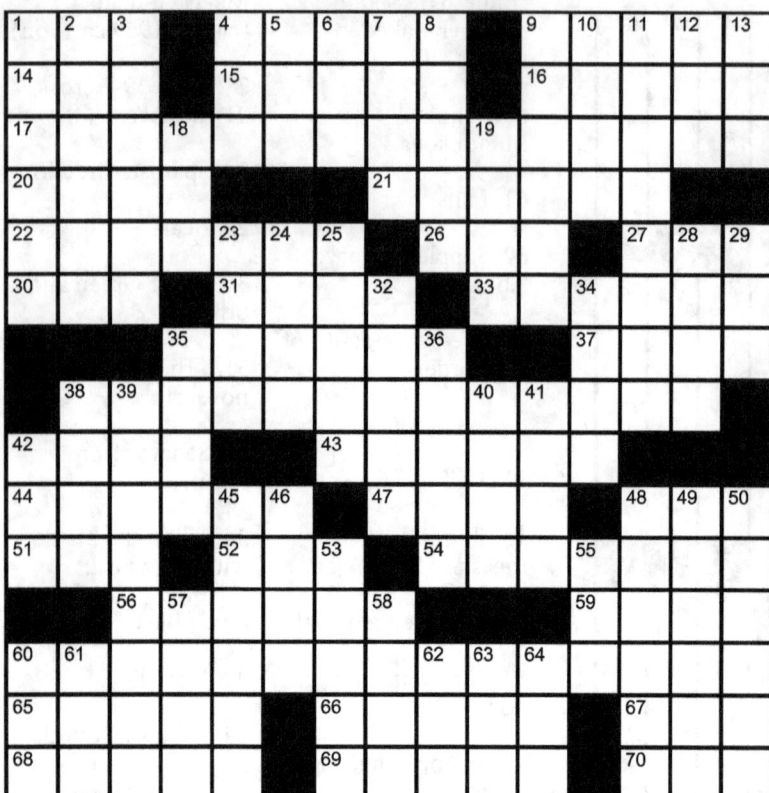

DOWN

1. Subject of a sailor's weather maxim

2. Title character in a Peter Hoeg best seller

3. Vance of "I Love Lucy"

4. Rock's ___ Soundsystem

5. We'll teach you to drink deep ___ you depart': Hamlet

6. What a H.S. dropout may get

7. Fill in the blank with this word: "" ___ teaches you when to be silent": Disraeli"

8. Mutual of ___

9. Stock ticker inventor

10. Whit

11. Western festival

12. Fill in the blank with this word: "___ el Amarna, Egypt"

13. Surgery sites, for short

18. TV's Cousin ___

19. Word ending meaning "foot"

23. Union member

24. Weightlifter's rep

25. Wiesbaden's state

28. Fill in the blank with this word: "___-Tass news agency"

29. Stowe girl

32. Wakens

34. Understands

35. Fill in the blank with this word: ""The

Bridge on the River ___""

36. Sounds that may be heard before bangs?

38. Six-stringed instrument

39. a purely biological unfolding of events involved in an organism changing gradually from a simple to a more complex level

40. Garage container

41. To laugh, to Lafayette

42. Group formed at C.C.N.Y. in 1910

45. More elegant

46. With 29-Down, central role on "Knots Landing"

48. Singer with the 1994 #1 hit "Bump N' Grind"

49. Child's attention-getting call to a parent

50. One of the Crusader states

53. Tetra- plus one

55. Fill in the blank with this word: ""The ___ Daba Honeymoon""

57. Trix alternative?

58. Fill in the blank with this word: ""No ___!""

60. Pulls a certain prank on, informally

61. Fill in the blank with this word: ""Are ___ pair?" ("Send in the Clowns" lyric)"

62. Photog's item

63. Ransom ___ Olds

64. Some nouns: Abbr.

ACROSS

1. Biblical ed.

4. Have a ___ stand on

9. Tissue: Prefix

14. Fill in the blank with this word: "Capitol-___ (music company)"

15. You should whip this ingredient before you top your Chantilly potatoes with it

16. Overindulgent parent, e.g.

17. Title for this puzzle

20. Use a knife

21. Sign for May Day babies

22. Gossipy group

26. Word modifier: Abbr.

27. Word with boss or

bull

30. PBS's '___ Can Cook'

31. Mardi Gras, e.g.: Abbr.

33. Fill in the blank with this word: "___ Field"

35. Limestone regions with deep fissures and sinkholes

37. Fill in the blank with this word: "___ : hello :: hooroo : goodbye"

38. Feature of 20- and 35-Across, forward and backward

42. Singer Turner

43. 13th-century king of Denmark

44. Like many a grandparent

47. Wrangle

48. Line score letters

51. West Bank grp.

52. Literary monogram

54. Pussyfooted

56. in gear (similar term)

59. Ohio native

60. California city by Joshua Tree National Park

65. Tubular pasta

66. Infection fighter

67. Rock's ___Lonely Boys

68. Leo with the 1977 #1 hit 'You Make Me Feel Like Dancing'

69. The English translation for the french word: Ègide

70. Warbler Sumac

PUZZLE 26

ACROSS

1. Fill in the blank with this word: "Director Gus Van ___"

5. Uplift spiritually

10. Fill in the blank with this word: "Co-___ (appropriates)"

14. Talon

15. Louvre, par exemple

16. Vardalos and Peeples

17. Vapory beginning

18. It's applied with a brush

20. Curve on the surface of a sphere

22. Satellite ___

23. Fill in the blank with this word: ""With this ring ___ wed""

24. As a welcome change

26. Water, e.g.: Abbr

28. Water pipes

30. Skin flicks

35. Kind of agent

38. Pool competitor's request

39. Mozart's "___ fan tutte"

40. Palm reader's reading

42. Yours: Fr.

43. The English translation for the french word: desceller

45. Have good intentions

47. Only reigning pope to write an autobiography

48. Some child-care center sites, for short

49. Year in Elizabeth I's reign

51. Museum with an Edward Hopper collection

55. Unexpected sports outcome

59. Fill in the blank with this word: "Costa Rica's ___ Peninsula"

61. Busts

62. Do a post-Challenger diagnosis

65. Writer Jaffe

66. Fill in the blank with this word: "Auvers-sur-___, last home of Vincent van Gogh"

67. Remove the dirt

68. You may bid on it

69. Fill in the blank with this word: "Children's author Ennis ___"

70. Val d'___, French ski resort

71. Zaire's Mobutu ___ Seko

DOWN

1. Winter wear

2. Even if, briefly

3. Belgian city or province

4. See 35-Across

5. Worker with a saving plan, for short

6. Like some ulcers

7. The writing ___ the wall

8. Kind of position

9. "Uh-huh"

10. Go ___ some length

11. Word of contempt

12. New Criticism poet Allen ___

13. Go on a vacation tour

19. Tickle, as one's interest

21. Fill in the blank with this word: "___ B'rith"

25. Spiny-rayed aquarium fish

27. Gettable

29. Vile

31. What public oddities often do

32. Must've been something ___'

33. Environmental sci.

34. Power tool brand

35. Support spec

36. Fill in the blank with this word: "___ soit qui mal y pense"

37. Fill in the blank with this word: ""Mi casa ___ casa""

41. Double or nothing, say?

44. We ___ please'

46. Shaving alternative

50. Winds

52. Mythical eponym of element #41

53. Writer Buchanan and others

54. Fill in the blank with this word: "Belgian violin virtuoso Eugene ___"

55. Wife, in legalese

56. Fill in the blank with this word: "___-dieu"

57. Writer's Market abbr.

58. Fill in the blank with this word: ""___ only""

60. Snick-or-___

63. You get one for a sac fly

64. Harpoon

PUZZLE 27

ACROSS

1. ___ Verde National Park

5. Bishop of Rome

9. Decrease

14. Intensifies, with "up"

15. Crown

16. Orange-red berrylike fruits

17. Grasslands

18. Pink, as a steak

19. Genealogy

20. Not easily destroyed

23. Remove salt from

24. Brown, e.g.

25. Arch

26. Dermatologist's concern

28. Appropriate

31. Going to the dogs, e.g.

34. Cheese on crackers

35. Civil aviation

36. A new industry

39. Red ink amount

40. Wine holder

41. Gets licked

42. "C'___ la vie!"

43. Quad building

44. Store convenience, for short

45. Propel, in a way

46. Pledge

49. That toward which you are inclined to feel dislike

54. Conclusion

55. Stalk

56. Anger

57. Host

58. Fungal spore sacs

59. Western blue flag, e.g.

60. A long time

61. Quaker's "you"

62. Actors

DOWN

1. French Sudan, today

2. Correct, as text

3. Digger

4. Judges

5. In part; in some degree; not wholly

6. Eyeball benders

7. Andean land

8. Board member, for short

9. Get there

10. Fool

11. Missing from the Marines, say

12. "Soap" family name

13. "Star Trek" rank: Abbr.

21. Game ragout

22. Beat

26. Geometrical solid

27. Farm call

28. "Hamlet" has five

29. Reduce, as expenses

30. 1992 Robin Williams movie

31. "Cast Away" setting

32. Couples

33. Acad.

34. Den denizen

35. Related by an isometry

37. Disdain

38. The "U" in UHF

43. Talking points?

44. Mame, for one

45. Basket material

46. Because

47. Beauty pageant wear

48. Hindu ascetics

49. Kind of store

50. Ancient Andean

51. ___ room

52. Luxurious

53. Home, informally

54. Effeminate

PUZZLE 28

ACROSS

1. Boring
5. Result of some plotting
10. Look sullen
14. Hokkaido native
15. Hindu queen
16. 100 cents
17. An ethical or moral code that applies more strictly to one group than to another
20. "Star Trek" rank: Abbr.
21. Abominable Snowman
22. Arise
23. "___ and the King of Siam"
24. Fabric
26. With all the bells and whistles
29. Andes capital
30. Caribbean, e.g.
33. Beasts of burden
34. Bartender's supply
35. Amigo
36. A forecast of the weather
40. Amniotic ___
41. Condos, e.g.
42. Sundae topper, perhaps
43. Addis Ababa's land: Abbr.
44. Frost-covered
45. Inclines
47. "What are the ___?"
48. Supergarb
49. Ribbon holder
52. Doctor Who villainess, with "the"
53. "Fancy that!"
56. Tubular passage of mucous membrane and muscle extending
60. Canine cry
61. Blood carrier
62. Charge
63. Obi, e.g.
64. Be exultant
65. ___ milk

DOWN

1. Commanded
2. Animal with a mane
3. The excretory opening at the end of the alimentary†canal.
4. Airline's home base
5. "The Power and the Glory" novelist
6. Bob Marley fan
7. The "A" of ABM
8. ___ green
9. Biddy
10. Inclined
11. "One of ___" (Willa Cather novel)
12. Language of Lahore
13. Gift on "The Bachelor"
18. Wildcat
19. Scratch up
23. Em, to Dorothy
24. Some brown-baggers
25. Arabic for "commander"
26. Search for water
27. A license for absence from a college or a religious house
28. Percolate
29. Like some goals
30. Extra
31. Artist's stand
32. ...
34. Cuts back
37. Runner's obstacle
38. "Idylls of the King" character
39. Bay
45. Tree that spreads laterally
46. "Beowulf," e.g.
47. ...lan
48. Bill of fare
49. "___ who?"
50. "Guilty," e.g.
51. Alternative to acrylics
52. Pink, as a steak
53. Vex, with "at"
54. "___ on Down the Road"
55. The "E" of B.P.O.E.
57. 40 winks
58. Anderson's "High ___"
59. Absorbed, as a cost

PUZZLE 29

ACROSS

1. It was tested on Bikini, 1954

6. Sternward

11. Infant fare

14. 'King Olaf" composer

15. Pepsi and Coke, but not 7UP

16. Bravo or Lobo

17. Expressed orally

19. Shelley praise

20. Unimaginative

21. Cell†division that produces reproductive†cells

23. Pub order

24. Cell†division that produces reproductive†cells

25. Clergyman's closetful

28. Taboo

32. '___ End"

33. One to hang with

34. Place for a massage

36. Thus

39. Like a stuffed shirt

41. It may be present in undercooked meat

42. Gherkin kin

43. Fencing Academy blade

44. Poisonous shrub, sometimes

45. Small amount

46. Cell substance

48. Computer architecture acronym

49. Workout spots

50. Toyota-made autos

53. ___-Wan Kenobi

55. Jamestown crop

57. Chart-topping country band

61. Christmas bulb, e.g.

62. Moderately sweet raised roll

64. Of a previous time

65. Drug from poppies

66. Drive out of bed

67. Like some ears

68. Musical interludes

69. Puts money in the pot

DOWN

1. Does a hatchet job on?

2. Use a paper towel

3. Shrek, e.g.

4. Patterned cotton fabric

5. One way to cook steak

6. The highest point

7. Word that leads to some starts?

8. One with an old school tie?

9. Ill-___ (doomed)

10. Place for a slogan

11. economic state of growth

12. Adjutants

13. Art of verse

18. Strip the blubber

22. Expressed wonderment, in a way

25. Cathedral nook

26. 'Post" or "light" leader

27. It goes to the winner

29. Transpire

30. Woman from Bethlehem

31. Earthenware pots

35. Highest-pitched woodwind

37. Rock music genre

38. Early carmaker

40. Columbus' reputed birthplace

47. Popeye's tattoo

49. 'Gibson†girl' (1867-1944)

50. Summer ermine

51. Welsh canine

52. Rifleman's aim improver

54. Seaport of Iraq

56. Singer Redding

57. Instruments of war

58. Adjoin

59. She's an inspiration

60. Tamandua's diet

63. Director's cry

PUZZLE 30

ACROSS

1. Ball field covering

5. "Check this out!"

9. High-hatter

13. Aroma

14. Blood of the gods

16. Bring on

17. Anniversary, e.g.

18. Nigerian monetary unit

19. ___-friendly

20. Harsh Athenian lawgiver

22. Position in a graded series

24. Cleave

26. Safari sight

27. Kind of first-aid pencil

30. Cousins of crunches

33. Two large muscles of the chest

35. Razor sharpener

37. www.yahoo.com, e.g.

38. Hackneyed

41. "Walking on Thin Ice" singer

42. Ancient Celtic priest

45. Medical exam

48. Overseas

51. Complains

52. Baffled

54. Banquets

55. accident or natural disaster

59. Spoonful, say

62. "God's Little ___"

63. Dostoyevsky novel, with "The"

65. Stronghold captured

66. Synagogue

67. Browning's Ben Ezra, e.g.

68. "I ___ you!"

69. Cozy and comfortable

70. Computer instructions

71. ___ probandi

DOWN

1. Mary in the White House

2. Jewish month

3. Service club and to promote world peace

4. Maxim

5. A.T.M. need

6. Heroin, slangily

7. Bake, as eggs

8. Ark contents

9. Prevent from entering

10. Not yet final, at law

11. Sundae topper, perhaps

12. European capital

15. Pie cuts, essentially

21. "I'm ___ you!"

23. Aardvark fare

25. Gossip

27. Tater

28. ___ cotta

29. "Wheels"

31. Collection of people or animals or vehicles moving ahead in more or less regular formation

32. Navigational aid

34. Back talk

36. Successful runners, for short

39. "___ will be done"

40. Young falcon or hawk

43. With anger

44. ___ any here know me?': King Lear

46. Blue books?

47. Pigment thickly so that brush

49. Buzzing

50. Fragrant Himalayan tree

53. Accused's need

55. 100-meter, e.g.

56. Bounce back, in a way

57. Jack-in-the-pulpit, e.g.

58. Arcing shots

60. Balsam used in perfumery

61. Aims

64. ___-tac-toe

PUZZLE 31

ACROSS

1. Wanted-poster letters

4. Peter who wrote "The Valachi Papers"

8. Unconscionably high interest

13. Pres., to the military

15. Vision: Prefix

16. Spasm

17. In days of knights?

19. "It's about time!"

20. Guy Lombardo hit of 1937 or Jimmy Dorsey hit of 1957

21. Fill in the blank with this word: "___ one-eighty"

23. Tony's portrayer on "NYPD Blue"

24. Look over again

25. Singer ___ King Cole

27. Hero in "Ulysses"

34. Yucat

37. Fill in the blank with this word: "Arkansas's ___ Mountains"

38. Name placeholder in govt. records

39. Fill in the blank with this word: ""The Seven Year ___""

40. "In & Out" star, 1997

41. Coal-rich German region

42. Nickname of 1954 home run leader Ted

43. Fill in the blank with this word: "Dr. ___ Hahn of "Grey's Anatomy""

44. Other: Fr.

45. Tommy Moe's specialty

48. C.I.A. : U.S. :: ___ : Soviet Union

49. Snifter filler

53. The muse of history

56. Fill in the blank with this word: "___ sister"

59. Funny little story

60. Updated, perhaps

62. Extinct duck-billed beast

64. Look-___ (twin)

65. Fill in the blank with this word: "Drive-___"

66. Yarn

67. Meaning of the name Eli

68. Fill in the blank with this word: "___

69. Ways around: Abbr.

DOWN

1. Make ___ for (support)

2. Natalia Makarova joined this company upon graduation from the Leningrad Choreographic School

3. Cavern, in poetry

4. To a greater extent

5. Bee: Prefix

6. Very little

7. Mozart's "Dove ___"

8. Tony-winning Hagen

9. "Fiddler on the Roof" setting

10. WWW addresses

11. Sub ___ (secretly)

12. The English translation for the french word: yÈti

14. often formed into braided loaves and glazed with eggs before baking

18. Steve ___ in "Family Matters"

22. Wrestling's ___ the Giant

26. Attempt

28. Now a 5-letter suffix, it was the Greek term for the type of community we call a city-state

29. Modern writer Cynthia

30. Maui neighbor

probandi"

31. Go ___ some length

32. Sheik ___ Abdel Rahman

33. The English translation for the french word: bourbe

34. Endangered Asian deer

35. Fill in the blank with this word: ""___ never work!""

36. Atlantic food fish

40. Maynard G. ___ of "The Many Loves of Dobie Gillis"

41. Propose

43. School subj.

44. That's ___!' (parental admonition)

46. Mardi Gras song that was a 1965 hit for the Dixie Cups

47. Victim of sun burn?

50. Big tournaments for university teams, informally

51. X-rated

52. Goddess of agriculture

53. To study intensely at the last minute for a test

54. Sir Peter ___, painter of British royalty

55. Yeah, man!'

57. Fill in the blank with this word: ""___ be in England...""

58. U-___ (German subway)

61. Like some races and hopes

63. R. & B.'s ___ Hill

PUZZLE 32

ACROSS

1. Bing Crosby's record label

4. Where Prince Philip was born

9. Malt liquor foams

14. Fill in the blank with this word: ""Just Another Girl on the ___" (1993 drama)"

15. Vast steppes make up a section of this mountain range that divides Europe & Asia

16. Trifles: Fr.

17. Be indifferent about

20. Dusseldorf donkey

21. The Big Apple's ___ Station

22. Hands out, as duties

26. Goes (for)

31. Fill in the blank with this word: "___ Offensive"

32. Schumacher of auto racing

34. Original "Playboy"

35. Some clouds

37. Wood sorrels

38. President's bird?

42. Fill in the blank with this word: "Actor ___ Phillips"

43. Work of Juvenal

44. World Cup shout

47. Young lady of Sp.

48. Year the Paris M

51. They may be half or full

53. Trinidad's Pitch Lake is a natural lake of this black substance

used to surface roads

55. The English translation for the french word: prison

57. Weapon in fencing

58. MAC

65. The U.N.'s Kofi ___

66. The English translation for the french word: ionique

67. Keats's "Ode on a Grecian ___"

68. Thin strips of wood

69. Walking ___ (happy)

70. ___ Precheck

DOWN

1. Patronize, as a restaurant

2. Game stick

3. The English translation for the french word: tÈmoigner

4. Fill in the blank with this word: "___ bono (for whose benefit?: Lat.)"

5. One of the Wright brothers, for short

6. Union activist Norma ___ Webster

7. Uproar

8. Wedding helper

9. One of three literary sisters

10. Small island

11. Whistler, at times

12. Telephone trio

13. Ukraine, e.g., formerly: Abbr.

18. Los : Spanish :: ___ : Italian

19. Small and insignificant

23. The English translation for the french word: larve

24. Movie whale

25. Staff connections

27. Prior to, old-style

28. Fill in the blank with this word: "Early film director Thomas H. ___"

29. Unscramble this word: lane

30. Yearbook signers: Abbr.

33. Tiny pests

35. Members of a raiding party

36. The English translation for the french word: interne

38. This small rodent whose name rhymes with mole is closely related to the lemming

39. The Shelters of Stone' writer

40. Rests

41. Fill in the blank with this word: ""That's a ___!""

42. Fill in the blank with this word: "Cambodia's ___ Nol"

45. Times going onto a secure site

46. Onetime Spanish queen and namesakes

48. Plan, as an itinerary

49. Unclogs

50. Workshop of Hephaestus

52. Old Buckeye State service station name

54. TV's "___ Ramsey"

56. Scientology's ___ Hubbard

58. Francis Poulenc's "Le ___ masqu

59. Transfer ___

60. Fill in the blank with this word: "1099-___ (tax form sent by a bank)"

61. Panama, e.g.

62. Santa ___, Calif.

63. Richard ___

64. Money machine mfr.

PUZZLE 33

ACROSS

1. WTO predecessor

5. TV's "___ and Greg"

11. Fill in the blank with this word: ""Polythene ___" (Beatles song)"

14. With 3-Down, way up for a downhiller

15. Like some relations

16. Fill in the blank with this word: "___ pro nobis"

17. "Later!"

19. Top part of a disguise

20. Sounds of doubt

21. a major road for any form of motor transport

23. Middle: Prefix

26. Fill in the blank with this word: "Electric ___"

28. Fill in the blank with this word: "___ gratia (in all kindness): Lat."

29. In a despicable way

31. Menacing look

33. Quarrel

34. Transverse rafter-joining timber

36. Classic Ernest L. Thayer poem

41. Greg of "You've Got Mail"

42. Scott Joplin's "Maple Leaf ___"

44. The English translation for the french word: revoir

47. Hand-held telescope

50. Hula ___

51. Wore away

52. "I love you, Juanita"

53. Selected athlete

56. Year of Bush's swearing-in

57. Year in the rule of Ethelred the Unready

58. People of much experience

64. WWII sphere

65. Swedish money

66. Supply-and-demand subj.

67. Stock units: Abbr.

68. Something to exercise in

69. Terza ___ (Italian verse form)

DOWN

1. Volkswagen model

2. The English translation for the french word: guichet automatique bancaire

3. Way of the East

4. Strictly in the style of

5. U.K. military medals

6. Publishing mogul, for short

7. Chopping part of a chopper

8. Oscar and Tony winner Mercedes

9. Husband: Fr.

10. Without ___ to stand on

11. One of the few words in the dictionary with 3 "W"s, it's a Native American get-together

12. The English translation for the french word: Ariane

13. Native Hungarian

18. Fill in the blank with this word: "Drive-___"

22. Weapon since 1952

23. Massachusetts' Cape ___

24. The "E" of N.E.A.: Abbr.

25. Thatching palm

26. Fill in the blank with this word: "___ Gonz"

27. What all the answers to should be clear?

30. The Party's Over' composer

31. Second-largest moon of Uranus

32. Ka ___ (Hawaii's South Cape)

35. Recording session need

37. Sounds like a broken record

38. Sue Grafton's '___ for Evidence'

39. Trapped like ___

40. Loc. of some devils

43. Mil. aide

44. "Don King: Only in America" star

45. Prehistoric stone tool

46. Swedish imports

48. Rooms full of sweaters?

49. Tarry

51. Unlawful firing?

54. Toll rds.

55. ___ The Magazine (bimonthly with 35+ million readers)

56. Zeppo, for one

59. Fill in the blank with this word: "February ___ (Groundhog Day)"

60. Tommy ___, Olympic skiing gold medalist

61. Motel freebie

62. Fill in the blank with this word: "CD-___"

63. Winter weather, in Edinburgh

PUZZLE 34

ACROSS

1. Fill in the blank with this word: ""Forever, ___" (1996 humor book)"

5. Wife in "8 Simple Rules for Dating My Teenage Daughter"

9. Per ___ ad astra (motto of the Royal Canadian Air Force)

14. Fill in the blank with this word: "Author L. Frank ___"

15. Ft. Worth's ___ Carter Museum

16. Von Trapp girl who's "sixteen going on seventeen"

17. Half of MCCIV

18. Unit of speed

19. Make ___ of it'

20. Departed a sheikdom?

23. The "D" in R&D: Abbr.

24. The English translation for the french word: attaquer

25. Syrupy treat

27. Big tournaments for university teams, informally

30. Fill in the blank with this word: "Al ___, 1984 Olympic gold medalist in the triple jump"

31. Unscramble this word: ybu

34. Monster slain by Perseus

36. Fill in the blank with this word: ""Mona ___""

39. Nile queen, informally

41. Popular Japanese beer

42. Fill in the blank with this word: "Broadway's ___ Simon Theatre"

43. Zodiacal delineation

44. Much may follow it

45. Fill in the blank with this word: "Conductor ___-Pekka Salonen"

46. Legislative grp.

48. Symbol seen on viola music

52. One of the brothers Grimm

55. Romance novelist ___ Glyn

59. Vitamin C source

60. Start of instructions for solving this puzzle

63. Rope with a loop

65. Part of A.A.U.W.: Abbr.

66. Kind of meeting in "O Brother, Where Art Thou?"

67. The English translation for the french word: boson

68. The Beatles' "Penny ___"

69. Fill in the blank with this word: "___ ease"

70. Position

71. Some mil. awards

72. Loch ___ monster

DOWN

1. The only known filoviruses are Marburg, which causes green monkey fever, & this one

2. Storms

3. Flubs

4. The Joy Luck Club' author

5. Secret doctrine

6. Wet nurse

7. Fill in the blank with this word: "___ de force"

8. Fill in the blank with this word: ""___ a Man" (Calder Willingham novel and play)"

9. Support group

10. Oysters ___ season

11. Freshen

12. St. Paul-to-Fargo rte.

13. Tylenol rival

21. Wite-Out manufacturer

22. Twain's ___ Joe

26. Singer Green with multiple Grammys

28. Ugandan tyrant Idi ___

29. Polar irregularity

31. Email letters

32. Eskimo knife

33. Polite acceptance

35. Tiddlywink, e.g.

37. Geometric figs.

38. Fill in the blank with this word: "___ Khan"

40. Sunfish or moonfish

41. Volleyball stat

47. Smooth over

49. What all the shaded answers have

50. Fill in the blank with this word: "England's Isle of ___"

51. Quarter-barrel

52. Fictional daddy

53. Fill in the blank with this word: "___-proof (easy to operate)"

54. Fungus, in Falmouth

56. Ronald Reagan's mother

57. Tracks

58. They're high in Manhattan

61. Transfer and messenger materials

62. Pres., to the military

64. Shelley's "___ Skylark"

PUZZLE 35

ACROSS

1. Tolstoy's "___ Fyodor Ivanovich"

5. Prefix with -meter

9. Sponsor of ads famous for nudity

13. Indian bovine

14. "Nadja" actress L

16. Make ___ for it

17. Zorro's marks

18. Welcome

19. You might give a speech by this

20. P

22. Resolve a longstanding disagreement

24. Realtor's specialty, for short

26. Miami-___

County

27. Miner's or caver's light generator

31. Hidden

34. The English translation for the french word: viscÈral

35. Wrap

37. Rhone feeder

39. Fill in the blank with this word: "Beethoven's "Archduke ___""

41. Muddle

43. Relative of "Oh, no!"

44. Finland, to the Finns

46. "It's about time!"

48. Fill in the blank with this word: "___ Rancho (suburb of

Albuquerque)"

49. Sounds that may be heard before bangs?

51. Old home decorations

53. General ___ chicken (Chinese menu item)

55. Tropical tree of the soapberry family

56. Can't continue

61. TV sports awards

64. Welcome, as a visitor / Try to make a date with

65. Old Houston hockey team

67. Slave to detail

68. Record-setting miler Jim

69. Went with

70. Where hops are dried

71. Sushi bar cupful

72. Soccer ___

73. Low-cost home loan corp.

DOWN

1. Theme of this puzzle

2. Look to ___ troublous world': 'Richard III'

3. Mental lapse

4. Showing signs of disuse

5. What you might do with hat in hand

6. Full of compassion

7. Pirate-fighting org

8. Like many old series, now

9. Theater area

10. Switch suffix

11. You might go for a spin in it

12. Working without ___

15. Fill in the blank with this word: "___ o' livin'"

21. Wings

23. Fill in the blank with this word: "___ fixe"

25. Wild ___

27. Ming's 7'6" and Bryant's 6'6", e.g.: Abbr.

28. East wind, in Greek myth

29. Poet ___ Van Duyn

30. Surveyors' maps

32. College student's filing

33. Pan Am rival, formerly

36. Black key

38. Wedding exchange

40. Minus

42. Where D.D.E. went to sch.

45. Fill in the blank with this word: "___ jure (by law)"

47. Something that takes its toll?: Abbr.

50. What a tragedy!'

52. Thinks ___ (disesteems)

54. Head of ___

56. Symbol on Irish euro coins

57. Fill in the blank with this word: ""For ___ sow...""

58. The English translation for the french word: stercoraire

59. Fill in the blank with this word: ""For here ___ go?""

60. Work with feet

62. Quaint affirmative

63. Young lady of Sp.

66. Vietnam War-era org.

PUZZLE 36

ACROSS

1. Dish on a stick: Var.

6. Was ___ hard on them?'

10. Short and disconnected: Abbr.

14. Fill in the blank with this word: ""Would you like to see ___?""

15. Window part

16. Tomb raider of film, ___ Croft

17. Fish-shaped musical instrument?

19. Shout when there's no cause for alarm?

20. Rescuer of Odysseus, in myth

21. Voting no

22. Irish runner Coghlan

24. Windhoek is its capital

26. Rhyming with clasp, it's a clasp for a door or lid that's fastened with a padlock

27. Night, to Nero

28. Poor custodian of time

31. We ___ please'

34. Weightlifter's rep

35. Site for techies

37. Small and insignificant

38. Wretched

39. Warm-hearted

40. The Naturist Society sponsors this type of recreation week, a chance for you to let it all hang out

41. You might take investing tips from this network's "On the Money" or "The Call"; Jon Stewart probably doesn't

42. Preceded, with 'to'

43. Middle school stage, commonly

45. Rock's ___ Jovi

46. The ___-Neisse Line

47. Fill in the blank with this word: "Brendan Behan's "___ Boy""

51. Lead-in to a questionable opinion

54. Fill in the blank with this word: "E. ___"

55. Yellow ___

56. Biblical verb

57. Shallow-water predator

60. Fill in the blank with this word: ""Young ___ Boone" (short-lived 1970s TV series)"

61. Prot. denomination

62. Vocalist Gorme

63. Bone: Prefix

64. Femme fatale in "The Carpetbaggers"

65. The plural of the word spy

DOWN

1. Summer camp shelter

2. Refrigerator brand

3. Twiggy broom

4. Switch ups?

5. Nightmare cause

6. Fill in the blank with this word: ""___ crime?""

7. Star of "Mr. Hulot's Holiday"

8. Fill in the blank with this word: "Costa Rica's ___ Peninsula"

9. "You don't say!"

10. T. Boone on a diet?

11. Fill in the blank with this word: "___-shanter"

12. Make ___ for it

13. Fill in the blank with this word: "___ Crunch"

18. Windows alternative

23. Visual way to communicate: Abbr.

25. Where to find a genie?

26. What's in carrots but not celery?

28. Words before a deadline

29. Writer Blyton

30. Noteworthy name in lens care

31. Get an ___ (ace)

32. Fill in the blank with this word: "___ to one's neck"

33. Monster slain by Perseus

34. Middle: Prefix

36. Group formed at C.C.N.Y. in 1910

38. Call-in radio show employee

42. Slow-moving primates

44. Teacher's deg.

45. Tree trunk

47. Fill in the blank with this word: "___ nova"

48. Letter before qoph

49. Where to live the high life?

50. Great ___

51. Fill in the blank with this word: ""___ Anything" ("Oliver!" song)"

52. With the intent

53. George Harrison's "___ It a Pity"

54. Crepe paper feature

58. Zogby poll partner

59. Under: Prefix

PUZZLE 37

ACROSS

1. Popular chocolate snack

8. Teleflora competitor

11. Radar, e.g.: Abbr.

14. Wheaties box candidate

15. The Everlys' "When Will ___ Loved"

16. Hand (out)

17. Joseph Conrad classic

19. Unscramble this word: alp

20. Newcastle upon ___, England

21. The English translation for the french word: imine

23. Green ___

27. It ends at Cairo

30. Zipped by

32. Ralph Vaughan Williams's "___ Symphony"

33. Top gun

34. The English translation for the french word: DSM

35. Fill in the blank with this word: "Amniotic ___"

38. Griddle order

42. Fill in the blank with this word: ""I'm ___""

43. Go to ___

44. On the verge of

45. Fill in the blank with this word: "___ so"

47. Waldenbooks competitor

48. repetition of information (silently or aloud) in order to keep it in short-term memory

52. Lend-___ Act

53. On display

54. Voyage V.I.P.

56. Elementary suffix

57. Minnie?

64. Potus #34

65. Taoism founder Lao-___

66. Some fed. govt. testing sites

67. Winding road shape

68. "Whoopee!"

69. McDonald's empire builder

DOWN

1. Morse T

2. Fill in the blank with this word: "Cardio : heart :: ___ : ear"

3. TV band

4. Time of legend

5. Repel, as an attack

6. Give it ___

7. Tie again, as a necktie

8. Fill in the blank with this word: ""Fee, ___, foe, fum""

9. USA alternative

10. Yen

11. Mid fifth-century date

12. Point : line :: line : ___

13. The English translation for the french word: lÈpreux

18. Old Testament book: Abbr.

22. Fill in the blank with this word: "___ Farrow, Mrs. Sinatra #3"

23. The English translation for the french word: poubelle

24. With no help from the U.S., Simon turned for aid to this Caribbean nation that threw out the French in 1804

25. Overseas diplomat in N.Y.C., say

26. Trailer org.?

28. Old capital of Romania

29. Royal fern

31. Fill in the blank with this word: ""___ paratus" (motto of the U.S. Coast Guard)"

34. Toxic spray

35. Tempur-Pedic competitor

36. Los ___, Calif.

37. Unborn twin

39. Years on end

40. Item for a hospital patient

41. The English translation for the french word: cÙte

45. Repugnant exclamation

46. Theme of Ecclesiastes

47. Faultfinder

48. Where to sign a credit card, e.g.

49. Israel's Barak and Olmert

50. Development developments

51. With a needle: Prefix

55. Rap's Salt-N-___

58. Wiretapping grp.

59. Napoleonic marshal

60. Litmus bluer: Abbr.

61. Fill in the blank with this word: ""A guy walks into a ___ "

62. Wall Street deal, in brief

63. Upper-left key

PUZZLE 38

ACROSS

1. Rival

5. Worked the soil

9. Like some legal proceedings

14. Guy Lombardo's "___ Lonely Trail"

15. The English translation for the french word: orle

16. Successor to Clement VIII

17. The English translation for the french word: foulque macroule

18. Unstable leptons

19. Snares

20. What Dr. Frankenstein tried to do?

23. On the ___

24. Fill in the blank with this word: "___ tuna"

25. .___

26. Mandela's land: Abbr.

27. Fictional county, locale of 20-Across

33. Water tester: Abbr.

34. U.K. military medals

35. Revealing, as a dress

38. What's spread on a spreadsheet

40. Bright, to Brecht

42. Sounds made by 36-Across

43. Herculon's fiber

46. Fill in the blank with this word: "El ___ (cheap cigar, in slang)"

49. Reactor overseer: Abbr.

50. Where rock's R.E.M. was formed

53. Rose ___ rose...'

55. Width measure

56. Weezer's music genre

57. Whaler's org.

58. a group of many islands in a large body of water

64. The English translation for the french word: malaire

66. Chevrolet model

67. Scientology's ___ Hubbard

68. Peace

69. Old song "Abdul Abulbul ___"

70. Central knob of a shield

71. Fill in the blank with this word: "___ Granada (old Spanish colony in the Americas)"

72. Softens in water, in a way

73. Turn down, with "on"

DOWN

1. Fill in the blank with this word: ""Veni, vidi, ___""

2. Fill in the blank with this word: ""Let's Get ___" (1973 #1 hit)"

3. Those, to Tom

4. Wicker material

5. Nickname for a good kisser

6. Word-of-mouth

7. Month preceding Rosh Hashanah

8. Two Women' director

9. Big Italian daily

10. Fill in the blank with this word: "___ Tamid (synagogue lamp)"

11. Tour de France activity

12. Montreal team

13. Term of address, in urban slang

21. Fill in the blank with this word: ""If ___ You" (#1 Alabama song)"

22. Poultry

27. Tokyo, once

28. Flame Queen ___ (famous gemstone)

29. 1980's TV twosome

30. Thumbs-up response

31. Winter Palace residents

32. See 44-Across: Abbr.

36. Jacquerie

37. Puccini soprano

39. Where the outboard motor goes

41. Yankee Maris, informally

44. Yep, that's clear'

45. Thomas Moore's "___ Ask the Hour"

47. Baby blues

48. Spy Mata ___

51. Peach ___

52. Amass

53. Hero of the 1997 best seller "Cold Mountain"

54. The ___ Adventure, at SeaWorld

59. Harness part

60. Fill in the blank with this word: ""___ the jackpot!""

61. First word of Virgil's "Aeneid"

62. Oodles

63. Lennon's in-laws

65. Yamaha product, briefly

PUZZLE 39

ACROSS

1. Hic, ___, hoc

5. Hat designed Lilly

10. Scale with five sharps: Abbr.

14. The item seen here, or the taste it might have if made with lemons

15. Football Hall-of-Famer ___ Hirsch

16. Sofer of soaps

17. Black box on "The Addams Family"?

19. "Qu

20. The Nittany Lions: Abbr.

21. 'Waiting for the Robert ___'

22. Travels like a flying squirrel

24. Nevada's state tree

25. Shook down

26. Without a cover at night

29. Most debonair

32. Tickle, as one's interest

33. Winter frosts

34. Red Sea vessel

36. Fill in the blank with this word: "Faulkner's "Requiem for ___""

37. Poetic feet

38. The female of the speces Equus caballus

39. Fill in the blank with this word: "Christine ___, "The Phantom of the Opera" heroine"

40. Wall St. deals

41. Medieval guild

42. Missouri site of the Scott Joplin Ragtime Festival

44. With 8-Across, world's oldest subway system?

45. Just ___ (very little)

46. Lacquered metalware

47. Fill in the blank with this word: ""___ your hearts faint": Deuteronomy 20:3"

50. Sondheim's "___ Like That"

51. Flash ___ (faddish assembly)

54. Sacred Buddhist mountain

55. Slowly entering

58. Balance parts

59. Fill in the blank with this word: "___ Hirsch of "Lords of Dogtown""

60. Two out of twenty?

61. Walking encyclopedia

62. With time to spare

63. Revenuer

DOWN

1. WWW address starter

2. Spa sounds

3. Dweeb

4. Whse. unit

5. Bring to bear

6. Venusian, e.g.

7. Tribe in Manitoba

8. Ad ___

9. Monocle

10. Fancy equine coif

11. This fermented honey-&-water beverage was a favorite of Chaucer's miller & of the god Thor

12. Suffix with utter

13. Sox foes

18. Video store category

23. Fill in the blank with this word: ""My Name Is Asher ___""

24. Fruit pastry

25. 1989 Tom Hanks film, with "The"

26. W.W. I plane

27. Fill in the blank with this word: ""Tony n' ___ Wedding" (theater hit)"

28. Waters: Lat.

29. Margaret Mead's "Coming of Age in ___"

30. Used

31. The Belvedere ___ (Vatican sculpture)

33. Career diplomat Philip

35. Withdraw gradually

37. Uncomfortable

41. Trinity member

43. Without further ___

44. Fill in the blank with this word: "___ Tunes"

46. Security that matures in a year or less, briefly

47. Run easily

48. Webzine

49. Singer/songwriter Vienna ___

50. Saudi Arabian province

51. Memory: Prefix

52. Fill in the blank with this word: "Cup ___ (hot drink, informally)"

53. Cap'n's mate

56. I ___ Rock'

57. TV's Cousin ___

PUZZLE 40

ACROSS

1. Waiting in the wings

6. Wild time

11. They're not part of the body: Abbr.

14. Swiss dish of grated and fried potatoes

15. Utah's ___ Mountains

16. Supermodel Carol

17. UV

20. Fill in the blank with this word: "Capitol-___ (music company)"

21. Former org. protecting depositors

22. Fill in the blank with this word: ""Let's ___""

23. Stage direction that means "alone"

25. Singer Harris

28. Unable to range freely

31. Venerated symbols

32. Fill in the blank with this word: "Deux : France :: ___ : Germany"

33. Eggs

34. Letters on a brandy bottle

35. The English translation for the french word: RAU

36. enables the President to take over the United States airwaves to warn the whole country of major catastrophic events

38. Range part: Abbr.

41. Fill in the blank with this word: "Cinch ___ (Hefty garbage bag brand)"

44. Two qtrs.

46. Fill in the blank with this word: "Dionne Warwick's "I ___ Little Prayer""

47. Crescent-shaped

50. Many a patient

52. Middle third of a famous motto

54. Big battery type

55. Fill in the blank with this word: ""___ nerve!""

56. Fill in the blank with this word: ""___ Room" (2001 children's book)"

58. Hall-of-Fame basketball coach Hank

61. Griddle order

65. Fill in the blank with this word: " ___ Aquarids (May meteor shower)"

66. Natalia Makarova joined this company upon graduation from the Leningrad Choreographic School

67. Mexico City daily

68. Was on the bottom?

69. Scandinavian coin with a hole in it

70. React violently, in a way

DOWN

1. Fill in the blank with this word: ""What's Hecuba to him ___ to Hecuba": Hamlet"

2. Linguist Chomsky

3. Simpson attorney?

4. You can bank on it

5. "Nonsense!"

6. *Shoot perfectly

7. Leslie Caron musical

8. What ___ surprise!'

9. Brit. money

10. Panama, e.g.

11. Sentence shortener

12. Zigzag, in a way

13. Victrola part

18. Unconscionably high interest

19. Jerome Kern tune "___ Forget"

24. Fill in the blank with this word: ""___ get it!" ("Aha!")"

26. Fill in the blank with this word: ""Yo! ___ Raps""

27. Hip-hop's ___ Def

28. Lao-___

29. Fill in the blank with this word: "___ Beach, Hawaii"

30. Early statistical software

37. Libido

38. Least sissyish

39. Tackle-to-mast rope on a ship

40. Scottish refusal

41. Roman sandal

42. Fill in the blank with this word: "___ Onassis, Jackie Kennedy's #2"

43. Round Table title: Abbr.

45. Hong Kong neighbor: Var.

46. The "S" of R.S.V.P.

47. Miss America Myerson and others

48. Fill in the blank with this word: ""Who Do ___ Kill?" (1992 movie)"

49. Charged

51. Hypnotist Franz

53. When to say "Feliz A

57. Use in great excess

59. With a "K" at the end, it's a wooden toy; without, a group of legislators who voted together

60. Full of compassion

62. Hateful org.

63. Russian space station

64. Corduroy ridge

PUZZLE 41

ACROSS

1. Fill in the blank with this word: "___-Alt-Del"

5. Savings acct. protector

9. Fill in the blank with this word: ""Young ___ Boone" (short-lived 1970s TV series)"

13. Fill in the blank with this word: "___ Valley Conference"

14. Like Mount Rushmore at night

16. Expy.

17. It's you ___'

18. Winner of seven Tonys in 1980

19. Tombstone name

20. [See circled letters]

23. Driver's opportunity

24. Yoko ___

25. Sue Grafton's '___ for Ricochet'

28. Ukraine, e.g., formerly: Abbr.

29. Bills

32. Pachacuti's people

34. Starting point of the Freedom Trail

36. Old Soviet secret police org.

39. The Equality State: Abbr.

40. W.W. II vessels

41. Double-H of magic

46. any of several crested Old World birds with a slender downward-curved bill

47. Passion

48. Fill in the blank with this word: "___-Deutschland"

51. 16 oz.

52. Veracruz Mrs.

54. Sault ___ Marie

56. What science fiction movie do taxes and amine bring to mind?

60. Wire-haired terrier of film

62. Kind of a drag

63. Vacant areas

64. Zoomed

65. Witching hour follower

66. Suffix with my-

67. Wool weights

68. Pas ___ (gentle ballet step)

69. D.C. group

DOWN

1. Takes over

2. Violent upheaval

3. More frosted

4. Windblown soil

5. "Old Gringo" author Carlos

6. Pen fluids

7. Russian gold medalist ___ Kulik

8. Of a tart fruit: Prefix

9. Call in the game Battleship

10. Lemon and melon, e.g.

11. This is ___' (broadcast tagline)

12. British verb ending

15. Somewhat astringent, as wine

21. Speed skater Apolo Anton ___

22. Prefix with sphere

26. Sinead O'Connor album 'Am ___ Your Girl?'

27. Workers need them: Abbr.

30. Vitamin a.k.a. riboflavin

31. Russian space program started in the 1960s

33. Year Otto I became king of the Lombards

34. Tupperware sound

35. Lymph ___

36. It's you! What a surprise!'

37. The English translation for the french word: prison

38. Proved valid, in a way

42. Fill in the blank with this word: "Dennis ___ and the Classics IV (1960s-'70s group)"

43. From __ Eternity'

44. It's my understanding that ...'

45. About 40 degrees, for N.Y.C.

48. Saved, in a way

49. Intervene

50. Characters in 'Romola' and 'The Gondoliers'

53. Washington's ___ Stage

55. Villainous Shakespearean roles

57. These, made by parliament, "are like cobwebs, which may catch small flies, but let wasps and hornets break through"

58. Fill in the blank with this word: ""James Joyce" author Leon ___"

59. Watch face

60. Where the outboard motor goes

61. Fill in the blank with this word: "___-Pitch"

PUZZLE 42

ACROSS

1. Voice mail prompt

5. Wry comic Mort

9. Fibula neighbor

14. Mideast's Gulf of ___

15. Shall I compare ___ to a summer's day?'

16. Fill in the blank with this word: ""I Still See ___" ("Paint Your Wagon" tune)"

17. Fill in the blank with this word: "___ occasion (never)"

18. Fill in the blank with this word: "Astronomy's ___ cloud"

19. Fill in the blank with this word: "___ latte"

20. "Dance the night away!"

23. Harvesting for fodder

24. It's in the back row, right of center

25. Tea time, perhaps

27. Swindler

32. Old Greek coins

36. Fill in the blank with this word: ""Morning Dance" band Spyro ___"

38. Rubaiyat' rhyme scheme

39. Snowflake or crystal shape

41. Susan who co-starred in "Five Easy Pieces"

43. Research facility: Abbr.

44. Oven ___

46. Std. on food labels

47. Triple-platinum 1982 album with the #1 hit "Africa"

49. Fill in the blank with this word: ""___ tale's best for winter": Shak."

51. Fill in the blank with this word: "___ tide"

53. Recovers, with "up"

58. "So?"

63. Talk show host Lake

64. The English translation for the french word: omettre

65. Fill in the blank with this word: "Catch ___"

66. 'Twenty Years After' character

67. Three-part ordeal for H.S. students

68. Fill in the blank with this word: "Auvers-sur-___, last home of Vincent van Gogh"

69. Over here...'

70. Explosives

71. Vintage vehicles

DOWN

1. Was emboldened

2. Fill in the blank with this word: "___ vincit amor"

3. Sluggo's comics pal

4. Mushroom variety

5. Like some traffic

6. Yes, matey! Sailors use this word to hail a ship, or to attract attention

7. a heron rookery

8. Stop working so hard

9. Some of Moby's music

10. Zoological wings

11. Fill in the blank with this word: ""That's ___""

12. Org. for the Denver Gold and Chicago Blitz

13. Writer's supply: Abbr.

21. Out of shape

22. Puerto ___

26. Fill in the blank with this word: "___ fruit"

28. Snoozes

29. Picasso's muse Dora ___

30. Opening run

31. Okinawa port

32. What ___?'

33. The center of the Czech Republic's wool industry, it looks like it needs to buy a vowel

34. Where hops are dried

35. Time-share unit

37. Word of disappointment

40. W.W. II-era G.I., e.g.

42. Slangy street greeting

45. Surveyor's assistant

48. Refuse to yield

50. With the situation thus

52. Fill in the blank with this word: ""___, of golden daffodils": Wordsworth"

54. Tryster's escape route, maybe

55. Fill in the blank with this word: "___, meenie, miney, mo"

56. Fill in the blank with this word: "___ Rizzo, Dustin Hoffman role"

57. Tart fruits

58. Fill in the blank with this word: ""That's a ___!""

59. Website statistic

60. Worms cries

61. W.B.A. stats

62. verb secure with a bitt

PUZZLE 43

ACROSS

1. Fill in the blank with this word: "___ helmet (safari wear)"

5. Move ___' (Curtis Mayfield song)

9. Unscramble this word: rgea

13. Here ___, there...' ('Old MacDonald' lyric)

14. Used a kitchen utensil

16. Fill in the blank with this word: "___ Vista"

17. Quip, part 3

20. Ordinal suffix

21. Linear

22. Singer in the 1958 movie "Go, Johnny, Go!"

23. Roman square?

24. What ___ tell you?'

25. They're up

33. They're mined in Virgin Valley

34. Fill in the blank with this word: ""From ___ shining ..." ("America the Beautiful" lyric)"

35. Fill in the blank with this word: "Dry ___"

36. Over there, poetically

37. What "p" may stand for

38. The English translation for the french word: bafouer

39. Either of two books of the Apocrypha: Abbr.

40. Former CNN show "Evans & ___"

41. Fill in the blank with this word: "___ a limb"

42. Memo about Stephen King's "Christine"?

45. Fill in the blank with this word: ""Saving Private ___""

46. TV pooch

47. The reddish wood of this member of the rose family is used by cabinet makers; the nut, by candy bar makers

50. Fill in the blank with this word: "___ Aarnio, innovative furniture designer"

52. Fill in the blank with this word: "Feather ___"

55. Stalemate

58. The Golden Age of Roman literature runs from Cicero to this "Art of Love" author

59. Matthew 4:10: "Get the hence, _____"

60. Service org. since 1882

61. Paris's ___ Gauche

62. Cause of some impulsive behavior, for short

63. Vase's handle

DOWN

1. Fill in the blank with this word: "___ the way"

2. Not orig.

3. Fill in the blank with this word: "Comedy Central's "___.0""

4. Fill in the blank with this word: "___ polloi"

5. Town near Oakland

6. TV's Nick at ___

7. School in La Jolla: Abbr.

8. Worshiper's seat

9. Reminiscent of the 1890s

10. Veteran journalist ___ Abel

11. Memo abbr.

12. Explorer John and others

15. Not stick to the path

18. Valley of the Kings sites

19. Go from ___ worse

23. Pinochle combo

24. Chase Field team

25. Fill in the blank with this word: "Charles ___, "Gaslight" star, 1944"

26. Strike ___ (what models do)

27. Fill in the blank with this word: "___ 500"

28. Square, in 1950s slang, indicated visually by a two-hand gesture

29. Singer Horne and others

30. Verdi baritone aria

31. This term for a mechanical human comes from a Czech word for "forced labor"

32. Trial figure

37. Spanish inns

38. Spanish explorer ___ Bautista de Anza

40. *Group with the 2000 #1 hit "It's Gonna Be Me"

41. Pope's "___ Solitude"

43. used in imitation gold jewelry

44. The English translation for the french word: commission

47. Omnia vincit ___

48. Fill in the blank with this word: "Dolly ___ of "Hello, Dolly!""

49. 1014, in history

50. Town line sign abbr.

51. Greenland base for many polar expeditions

52. Windfall

53. Whacks

54. Chad's place

56. Sports org.

57. Longtime Chicago Bears coach

PUZZLE 44

ACROSS

1. Those Animals Frighten Me! : Ailurophobia

5. Wholly absorbed

9. The ___ Adventure, at SeaWorld

14. Defeat

15. Thought: Prefix

16. More modern

17. Fill in the blank with this word: "___ Polo of "Meet the Fockers""

18. Underworld figure

19. Indian ___

20. Sidestroke features

23. Capable, slangily

24. Guy Lombardo's "___ Lonely Trail"

25. Wisc. clock setting

28. Gets the better of

32. Lost interest in, in a way

34. Prefix with sac or duct

35. Finish this popular saying: "Hard work never did anyone any_____."

37. Neil Diamond's "___ Said"

38. Boeing plane

43. The Sopranos' actor Robert

44. Filmmakers: Joel & Ethan

45. One in the charge of un instituteur

46. Major League player before moving to S.F.

50. Less ornate

52. Listening to the facts spouted by the just-arrived talking antelope, I couldn't believe what the ____ ____ ____

53. Fill in the blank with this word: ""___ to please""

55. Topper

56. Nebulous stuff

62. "Rocky" score composer

64. Very little

65. They, in Italy

66. Fill in the blank with this word: ""Maid of Athens, ___ part": Byron"

67. Singer Simone

68. Proofs of purchase: Abbr.

69. Poet William Butler ___

70. Room to swing ___

71. Unscramble this word: shut

DOWN

1. Shipping units: Abbr.

2. Smart-___

3. TV actress Spelling

4. Obviously'

5. Ravioli fillings

6. Sixth Jewish month

7. Which liquid product featured the hit Like A Prayer in its advertisement?

8. Started liking

9. Vehicle with caterpillar treads

10. Fill in the blank with this word: ""What the ___!""

11. Amazes a horror film director?

12. Give __ break!'

13. Keats's "Ode on a Grecian ___"

21. Winningest southpaw in major-league history

22. Terre Haute sch.

26. Mother of Dionysus

27. Hair braider, e.g.

28. Some clinic work

29. Fill in the blank with this word:

"Crabtree & ___, purveyor of skin care products"

30. Like an old English coin worth 21 shillings

31. Indian poet ___ Aurobindo

33. Fill in the blank with this word: ""Oh wad some power the giftie ___ us": Burns"

36. Late 11th-century year

39. Verdi's "___ tu"

40. Fill in the blank with this word: ""You're the ___ " (Cole Porter classic)"

41. Berated loudly

42. These gastropods are sometimes fed aromatic herbs to give them a special savor

47. Factor in a hotel rating

48. Scottish refusal

49. European capital

51. Goof-offs

54. Old Connecticut whaling town

57. Wrapper abbr

58. Lake ___, head of the Blue Nile

59. Yiddish writer Sholem

60. V preceder

61. Mile or kilometer: Abbr.

62. Six-time All-Star third baseman of the 1970s Dodgers

63. Silver ___

PUZZLE 45

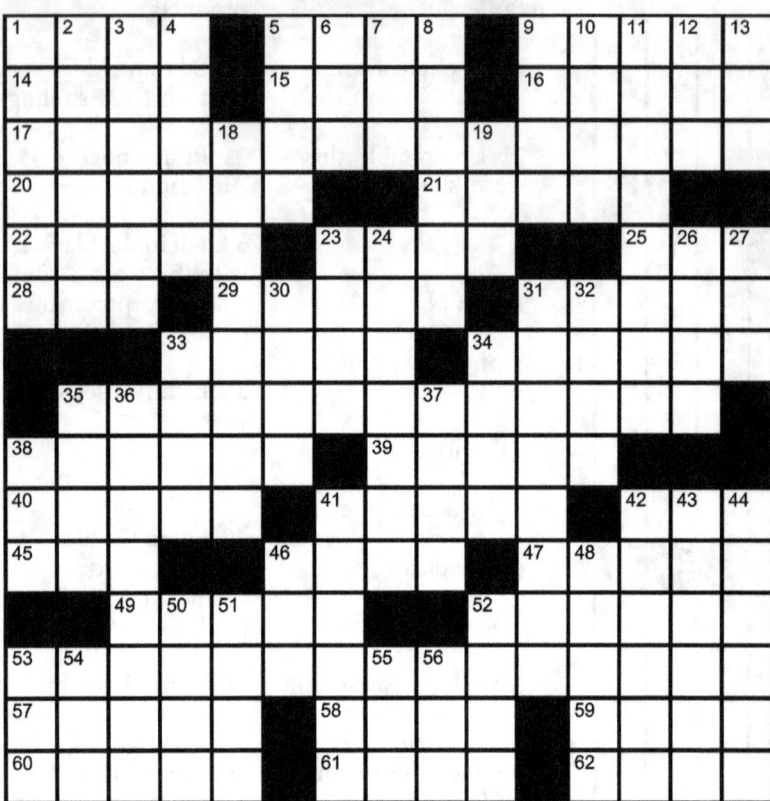

ACROSS

1. Woodstock gear

5. Intestinal parts

9. Finish this popular saying: "Little things please little_____."

14. Fill in the blank with this word: "Baby ___"

15. Wyo. neighbor

16. Whoops

17. Just the pits

20. Soothe, in a way, as a burn

21. Fill in the blank with this word: "___ Coyote"

22. Sixth-century Chinese dynasty

23. Some TV spots, briefly

25. Wellness grp.

28. Treebeard, e.g.

29. Modern writer Cynthia

31. Samuel Gompers's org., informally

33. Red Sox Hall-of-Famer Bobby

34. Grammy-winning 1996 Beck album

35. Principal principles

38. The English translation for the french word: nain

39. Composer who studied with Humperdinck

40. Swift as ___

41. TV's "The George & ___ Show"

42. Magic, on scoreboards

45. Wish

46. Long dist.

47. Titter

49. Firefighter Red

52. What the majority of elements are

53. Central Vermont, e.g.

57. Sunfish

58. Spanish ones

59. Unscramble this word: gang

60. Short hair, to Burns

61. Sew up

62. Villa d'___

DOWN

1. The English translation for the french word: arable

2. Pillowcase material

3. W.W. II naval craft

4. William ___, longtime editor of The New Yorker

5. Put ___ appearance

6. Fill in the blank with this word: "___ Cayes, Haiti"

7. Withdraw

8. Onetime Caribbean native

9. Year the emperor Frederick II died

10. Latin hymn "Dies ___"

11. Like some golf courses

12. Society newcomer

13. Wind dir.

18. Well-meaning sort

19. Trash

23. Weaver's bobbin

24. Like doodles

26. Some 6-Down curators: Abbr.

27. West Coast brew, for short

30. Time: Ger.

31. Worships

32. Tumbled

33. Bring (up) from the past

34. Mourning Becomes Electra' brother

35. Funny ___, 2003 Derby winner

36. West German Chancellor, 1949-63

37. Shakespearean king

38. Fill in the blank with this word: "___ Day"

41. Open-air room

42. The Wilkeses' neighbors

43. Yield

44. Writer of "Gil Blas"

46. Fill in the blank with this word: "Debussy's "Air de ___""

48. Story, in France

50. Strands of biology

51. Fill in the blank with this word: ""Look ___" (1975 #1 R & B hit)"

52. Yellow squirt?

53. Comic

54. Horatio! ___ do forget myself': Hamlet

55. Wichita-to-Omaha dir.

56. Suffix with ball or bass

PUZZLE 46

ACROSS

1. California Indian tribe: Var.

6. Toy gun shot

9. W.E. B. Du Bois & others founded this organization on Lincoln's 100th birthday in 1909

14. Fill in the blank with this word: ""Coffee ___?""

15. The Era of ___ (1964-74 Notre Dame football)

16. Grant's first secretary of state ___ Washburne

17. Smart guys?

18. It's found in a runoff

20. Covered stadium that's off-limits to bands?

22. Wrath

23. Water

24. Fill in the blank with this word: "Confit d'___ (potted goose)"

27. Tiffany showroom?

33. Most miserable hour that ___ time saw': Lady Capulet

34. Fill in the blank with this word: "___ Bank (U.S. loan guarantor)"

35. Wintry forecast

36. Sodium hydroxide, to chemists

38. Tower supports

41. Greenland base for many polar expeditions

42. Green ___

44. The English translation for the french word: roseau

46. Taxonomic suffix

47. Silver State boogie band autopsy expert

51. What's more

52. Something left of center?

53. Fill in the blank with this word: "___ show"

54. Depict part of the periodic table?

60. Assistance from a tall librarian?

63. Fill in the blank with this word: "___ grabs"

64. Fill in the blank with this word: "___ Beach (town near San Luis Obispo)"

65. You're killing me,' textually

66. French explorer La ___

67. Untighten

68. You're in balance if you know it's the complement of "yang"

69. Ten Commandments word

DOWN

1. The English translation for the french word: guËde

2. Fill in the blank with this word: "Comic strip "___ & Janis""

3. World War II weapon

4. marked by intense agitation or emotion

5. Bingeing

6. Vamp Theda

7. The English translation for the french word: hurler

8. Greeted, with "to"

9. Maternity ward figures

10. Zoological wings

11. Small island

12. Revolutionary Guevara

13. Water filter brand

19. Fill in the blank with this word: "___ one"

21. Successor to Clement VIII

24. Small bridge limit, maybe

25. Imagine

26. Williams in the water

27. Stanza alternative

28. The English translation for the french word: paonne

29. Fill in the blank with this word: ""I Get ___" (Beach Boys hit)"

30. White lie

31. German currency, informally

32. Grand ___ Opry

37. Wellness grp.

39. Fill in the blank with this word: "___ room"

40. You'll find the Kyongbok Palace at the foot of Mount Pugak in this South Korean capital

43. Uncle Remus character

45. Writer on Mulberry Street

48. Sunfish or moonfish

49. She won an Emmy playing Miss Jean

50. Tuba sound

54. Longtime Vicki Lawrence character

55. Wells's oppressed race

56. One-named singer/actress

57. Black key

58. Longtime Steelers coach Chuck

59. Waste allowance of old

60. White Sulphur ___, W. Va.: Abbr.

61. '___ you to horse': Macbeth

62. What you might hear halting speech in, for short

PUZZLE 47

ACROSS

1. Ziegfeld Follies designer

5. Tiny battery

9. Muslim judge

14. Honor for the A-team?

16. French peers

17. Fight imaginary foes

19. Southern Conference sch.

20. Warbler Sumac

21. Winter hrs. in Bermuda

22. Ali G portrayer ___ Baron Cohen

24. Subject to disproval

29. Department of eastern France

30. Former Georgia senator Sam

31. Old New Yorker cartoonist Gardner ___

32. Plotz

35. What you might buy a Gucci bag in?

36. Votes overseas

37. Opera first performed in 1762

40. Speak

41. True-crime TV series

42. Single-celled organism

43. Fill in the blank with this word: ""...___ thousand times...""

44. Peak near Neuch

45. Nav. leader

46. Picasso's "private muse"

48. Takes five

51. What le gendarme enforces

52. Tin ___

53. Team on which Larry Bird played, on scoreboards

55. Ice cream flavor #4

61. Seven-time All-Star pitcher Dave

62. Avoid having to wait for a table, maybe

63. Kafka character Gregor ___

64. Caught some rays

65. Fill in the blank with this word: "___ fixe"

DOWN

1. Ron Howard media satire

2. Site of Germany's surrender in W.W. II

3. Yo, she was Adrian

4. Treebeard, e.g.

5. The Player' director

6. Sony subsidiary

7. Fill in the blank with this word: ""___ was saying Ö""

8. When some stores open

9. Confines

10. Unsettled feeling

11. Fill in the blank with this word: "___ Kan"

12. Taken ___

13. Works for an ed.

15. Greet

18. Tony- and Emmy-winning actress Blythe

23. Seat fixer

24. a river that rises in Russia near Smolensk and flowing south through Belarus and Ukraine to empty into the Black Sea

25. East wind, in Greek myth

26. U.C.L.A. player

27. Quality camera

28. Slowly merged (into)

32. Discovered

33. Fill in the blank with this word:

"___-Detoo"

34. Alas...'

35. Treatment for Parkinson's

36. What your nose knows

38. The English translation for the french word: Èclair

39. The English translation for the french word: immersion

44. The name of this one-celled protozoan comes from the Greek for "change"

45. Ready for shipping

47. Where Jean-Claude Killy practiced

49. 'Vette rival

50. Pedestal support

52. Trix alternative?

54. Fill in the blank with this word: "___ terrier"

55. They're not part of the body: Abbr.

56. Fill in the blank with this word: "___ Aquarids (May meteor shower)"

57. Mr. ___ (absent-minded Milne character)

58. Ward workers, for short

59. Fill in the blank with this word: ""There is no ___ team""

60. Mai ___

PUZZLE 48

ACROSS

1. Fill in the blank with this word: "___ 10"

5. Unctuous flattery

10. Songs for one

14. High-tech transmission

15. Star quality

16. Warm-up for the college-bound

17. Singer/songwriter who received a 7-Down (2002, 2004-06, 2008)

19. Voice of America org.

20. Tree with double-toothed leaves and durable wood

21. *Sign to look elsewhere

23. Bazooka Joe's working peeper

25. Pancreatic enzyme

26. Fill in the blank with this word: "___ apso (dog)"

29. Annoying types

33. Fill in the blank with this word: ""___, I do believe I failed you" (opening of a 1998 hit)"

34. New York Senator

35. Verb type: Abbr.

38. She wed George Washington

41. See 85-Across

42. Some sorority women

43. Smutty

44. Owned (up)

45. 1988 Olympic track star, informally

46. It's ___ Kiss' (alternate name of 'The Shoop Shoop Song')

49. South Asians speak it

51. Suggests

55. Like supermarkets, theaters and planes

59. Fill in the blank with this word: "___ podrida"

60. Some radio productions

62. This member of the parsley family with a 4-letter name is a main flavoring agent in pickles

63. Understanding

64. Rat

65. Zaire's Mobutu ___ Seko

66. What an unrequited lover carries

67. Vous ___ ici'

DOWN

1. You have the right to do this regarding arms, but your arms will be this without sleeves

2. Fiery

3. The Shirelles' "Mama ___"

4. Has a flair for

5. Squalid

6. R & B singer Brian

7. Greek goddess Athena ___

8. They're caught on beaches

9. Fill in the blank with this word: "___ Mary's (L.A. college)"

10. Rejects rudely

11. Direction to an alternative musical passage

12. Flock members

13. Fill in the blank with this word: ""May ___ your order?""

18. Intestinal parts

22. Fill in the blank with this word: "Computer ___"

24. Slips by

26. Jesus, __ of God

27. Four hours on the job, perhaps

28. They're what a pompous person "puts on"

30. ___ 300 (short-lived Apple laptop)

31. Fill in the blank with this word: "___ 2.0, Bill Gates's house"

32. Waiting area announcements, briefly

34. Three of these make an O

35. Thought: Prefix

36. Actress Gilbert of "Roseanne"

37. Work over

39. Thermonuclear experiment of the '50s

40. Mistreatment

44. Windup

45. Savings acct. protector

46. Ubiquitous players

47. Fill in the blank with this word: ""And that's ___" ("Believe you me")"

48. Forest ___

50. With 6- and 22-Across, noted 19th-century writer

52. USMC rank

53. Novice: Var.

54. Sheik ___ Abdel Rahman

56. Would-be J.D.'s hurdle

57. Woman's name suffix

58. U.K. military medals

61. Western N.C.A.A. powerhouse

PUZZLE 49

ACROSS

1. Texter's 'Alternatively ...'

5. V preceder

9. Philippines' highest peak: Abbr.

14. Unsuitable for bluebloods, in England

15. Fill in the blank with this word: "___ out (manages)"

16. Where to get down

17. Barry B. Longyear novella that won Hugo and Nebula awards

19. McDonald's: "We love to see you ____"

20. any of several crested Old World birds with a slender downward-curved bill

21. They may play first

23. Vandal

24. Prefix with foil or phobia

25. Like some covers

29. Supplants

33. Three ___ match

34. Small mouthlike aperture

36. Novice: Var.

37. Fill in the blank with this word: "Father ___ Sarducci, longtime "S.N.L." character"

39. Fill in the blank with this word: ""It ___; be not afraid" (words of Jesus): 2 wds.'

40. Actor Sam

41. Bears: Lat.

42. Peak on the eastern edge of Yosemite Natl. Park

44. You could've left some of that out'

45. Alice who won an Emmy for "Bewitched"

47. Kansas town famous in railroad history

49. Terrell who sang with Marvin Gaye

51. Wooed very well

52. Greasy kid stuff

55. Kind of smoothie

59. Wee one

60. Helps for a time

63. Temple architectural features

64. German "genuine"

65. Fasten on

66. Yucky

67. TV/radio host John

68. Teen activist org.

DOWN

1. Small and insignificant

2. Up ___ good

3. Fill in the blank with this word: "___ cat"

4. Snorts of disdain

5. Patches again

6. Word with bum or bunny

7. Fill in the blank with this word: "Big ___ Conference"

8. Words on a perishable's container

9. Salutation abbreviation

10. Play breaks

11. Take ___ view of

12. Cosmos star

13. Opera conductor Daniel ___

18. It's not just me?'

22. Together, to Toscanini

24. Sonata maker

25. Get steamy

26. Toughen, as to hardship

27. Nancy's opposite number, once

28. Zwei x vier

30. Photographer Herb

31. TV teaser

32. Fill in the blank with this word: "___-law"

35. Would-be J.D.'s hurdle

38. Fink

40. University of Virginia players, familiarly

42. Fill in the blank with this word: ""I must have missed the ___""

43. Noted children's book illustrator (one of six "middle C" people in this puzzle)

46. TV cop Chris

48. They're worth their weight in gold

50. Some church music

52. Scottish seaport known for its single-malt Scotch

53. Suffix for a collection

54. Wagons-___ (sleeping cars, abroad)

56. Part of the eye

57. Sent back: Abbr.

58. Well-___

61. Italian ___

62. The N.L. doesn't allow them

PUZZLE 50

ACROSS

1. Shoe part

5. With 52-Across, what angels pray for

9. When a snake sticks out its tongue, it's not tasting but using this sense

14. Fill in the blank with this word: ""And ___ word from our sponsor""

15. In ___ of (replacing)

16. Keep an ___ the ground

17. Women with aur

18. William Saroyan's "My Name Is ___"

19. Well-balanced person?

20. Playwright in rare form?

23. Passed on by oral tradition

24. Indian poet ___ Aurobindo

25. Jazz phrase

29. Twosome

31. Like almost all TV shows nowadays

33. Org. for mom-and-pop stores

36. Sticker letters

38. Esther with home-maid roles

39. Singles bar acquisition

43. Russia's ___ Republic

44. Information source

45. Fill in the blank with this word: "Electric ___"

46. More hard-fought

49. Naut. direction

51. The English translation for the french word: fougËre

52. Fill in the blank with this word: "___ green"

54. London insurance giant

58. Baker's quote from "Romeo and Juliet"?

61. Blowing away

64. Willow variety

65. Fill in the blank with this word: "___-eyed"

66. Mediterranean resort island

67. Steals, old-style

68. Lived ___

(celebrated)

69. Win by ___

70. Word to a tabby

71. Snoozes

DOWN

1. Wordless song: Abbr.

2. Instruction for casual dress

3. Used a broom

4. Best Director of 1992 and 2004

5. Thrown together

6. Surrealist Joan

7. Talk show host Hannity and others

8. Arm bones

9. Unscramble this word: lsle

10. Stereotypical starting job assignment at a corporation

11. Tarzan creator's monogram

12. Xerox setting: Abbr

13. Mauna ___ volcano

21. ___ Games

22. Fill in the blank with this word: ""There is no ___ team""

26. Fill in the blanks with these two words: ""Well, ___!""

27. Fill in the blank with this word: "Det. Axel ___ of "Beverly Hills Cop""

28. Fill in the blank with this word: ""___ Jacques" (children's song)"

30. Be an utter bore?

32. apt to break into small fragments or disintegrate

33. Workers on duty

34. Run counter to

35. Unscramble this word: laret

37. Fill in the blank with this word: "___ Park, home for the Pittsburgh Pirates"

40. Salary

41. Workup locales: Abbr.

42. Last's correlative

47. Something left of center?

48. Two of the three gifts of the Magi

50. Knocks off

53. Where the last flight ends?

55. Where the United Nations' setup was discussed

56. Classic Marx Brothers flick

57. You can take them in stride

59. "Sleeping" sensation

60. Blood: Prefix

61. Fill in the blank with this word: ""The ___ Daba Honeymoon""

62. ___ America (Chicago-based superstation)

63. Fill in the blank with this word: "___ Jima"

PUZZLE 51

ACROSS

1. Naut. law enforcers

5. Washing jobs

10. Physics Nobelist Simon van der ___

14. Word with strings or horns

15. Gluck's "___ ed Euridice"

16. Fill in the blank with this word: "___ David"

17. Yield, as interest

18. 58-Across's first mate

19. Lit ___ (college course, slangily)

20. Fully, in a way

23. Fill in the blank with this word: "Caput ___ syndrome (arm problem)"

24. It's whipped to make mousse

25. Wrath

27. Fill in the blank with this word: "Debussy's "Air de ___""

28. Year in Claudius's reign

31. Pop's ___ Brothers

33. Ladles

37. Fill in the blank with this word: "___ even keel: 2 wds."

38. Bell ringer

41. Peach ___

42. Intelligently planned progress

43. Prefix with ribonucleic

45. Fill in the blank with this word: "___ cit."

46. Fill in the blank with this word: ""Humanum ___ errare""

49. Fill in the blank with this word: "Farmer's ___"

50. Toodles!'

54. Suggests

56. Declines, as a plumber?

60. New Criticism poet Allen ___

61. Lunar valley

62. Fill in the blank with this word: ""___ Simple Man" (#1 Ricky Van Shelton song)"

63. Memory: Prefix

64. Fill in the blank with this word: ""Let ___ Cake""

65. The English translation for the french word: scannÈriser

66. You never had ___ good!'

67. Father-and-daughter Hollywood duo

68. Fill in the blank with this word: ""___ Rebel" (1962 #1 hit)"

DOWN

1. utilizable (similar term)

2. Would you like me to?'

3. The English translation for the french word: couronne solaire

4. Salami choice

5. Finish this popular saying: "He who hesitates is_____."

6. Fill in the blank with this word: ""For here ___ go?""

7. Make ___ (mug)

8. Removal of restrictions, informally

9. Pulitzer-winning Ferber title

10. Turn-of-the-century year in King John's reign

11. Salary

12. Expatriate

13. Sch. assignment

21. Fill in the blank with this word: "___ of the past"

22. Fill in the blank with this word: "___ Ranch (former Western White House)"

26. Shingle abbr.

29. Fill in the blank with this word: "___ a Spell on You" (classic 1956 Screamin' Jay Hawkins song)"

30. Fill in the blank with this word: "___ dixit"

32. Suffix with psych-

33. Bond villain in "Moonraker"

34. What you might hear halting speech in, for short

35. Cambodian money

36. Unexciting

38. Unlikely to offend

39. Hard rubbers

40. The English translation for the french word: niche

41. Wyo. is on it in the summer

44. This puzzle's theme, phonetically

46. Intertwine

47. Plant pores

48. South African tongue

51. Comparable to a rose?

52. Cheat on

53. One who hears "You've got mail"

55. Words of longing

57. Little ___

58. Josephine Tey investigator ___ Grant

59. Trawling equipment

60. You could've left some of that out'

PUZZLE 52

ACROSS

1. Sheet of matted wool

5. Midwest and Plains states, e.g.

10. Fill in the blank with this word: "___ fruit"

14. Vitellius succeeded him

15. Moray catcher

16. Two-word airline name

17. Wriggler

18. Vigorous exercise system

19. Dos minus dos

20. Beer sources for genteel guests?

23. Dogpatch possessive

24. Ready already

25. Fill in the blank with this word: "1980 Peace Nobelist ___ P"

28. Fill in the blank with this word: ""Are you calling me ___?""

30. Putting all the poker chips in the pot, maybe

31. Spanish wine town

32. Underground org.

35. Neural network

36. Fill in the blank with this word: "Dear ___ Madam ..."

37. With 13-Down, "super power" glasses

38. Supermarket with a red oval logo

39. Tuckered out

40. Rock's Burdon and Clapton

41. Paris's ___-de-Medecine

42. Winds up

43. Fill in the blank with this word: "___ honorable (formal apology)"

46. Yes, matey! Sailors use this word to hail a ship, or to attract attention

47. Cadets, eventually

52. The Swedes who settled in Delaware in 1638 were the 1st in America to build cabins made of these

53. of or relating to yoga

54. Ja and da

56. Russia's ___ Mountains

57. What Spanish athletes go for at the Olympics

58. Tennyson's "immemorial ___"

59. Withered

60. Drops off

61. Cub #21 of the 1990s-2000s

DOWN

1. What a violinist may take on stage, in two different senses

2. Surmounting

3. Fill in the blank with this word: ""Comin' ___ the Rye""

4. "The Electric Kool-Aid Acid Test" author

5. The English translation for the french word: hÈtÈro

6. Wise up

7. Yankee or Angel, for short

8. Biblical peak

9. Less spirited

10. Catcher in the Wry' writer

11. Unaccompanied part songs

12. Living ___

13. Weird Al Yankovic's "___ on Jeopardy"

21. Rear

22. Jason who sang "I'm Yours," 2008

25. Dugout shelter

26. Golfer's challenge

27. Make ___ check

28. Where to live the high life?

29. Scientology's ___ Hubbard

31. Well-intentioned girl of rhyme?

32. "Trinity" novelist

33. This nutmeg spice has an aroma reminiscent of cinnamon & pepper

34. The Swiss Family Robinson' writer

36. Like some vamps

37. Producer of a piercing look

39. Fill in the blank with this word: ""___ Live," 1992 multiplatinum album"

40. Workplace fairness agcy

41. Strand, in a way

42. Spanish boys

43. Unbeatable mark

44. Fill in the blank with this word: "Actress Mary Tyler ___"

45. Sir Edward who composed "Pomp and Circumstance"

46. Like a house ___

48. With 121-Across, part of an afternoon repast

49. Hair line

50. Realtor's specialty, for short

51. Theological schools: Abbr.

55. What's funded by FICA, for short

PUZZLE 53

ACROSS

1. Baby's mind, e.g.

11. Some computers

15. Incitements

16. Fill in the blank with this word: "___-food industry"

17. Where legislators pass the time?

19. Work well together

20. Warp

21. Town on the Tappan Zee

22. Wakens

23. News exec Roger

24. Spanish skating figures

27. Something left of center?

28. The English translation for the french word: ravir

29. the earth from his flesh

32. Longtime essayist for The New Yorker

35. Stupidly silent

39. Want-ad letters

40. The Pointer Sisters' "___ Excited"

41. Proffer bait

42. Fill in the blank with this word: "Chinese author ___ Yutang"

44. Exceptional rating

45. Ancient Greek tongue: Var.

48. Career diplomat Philip

51. Without dissent

52. Standoffish

53. Scottish singletons

57. Famous landing site

60. They say piv while stealing havarti & squitt squitt if they get into the provolone

61. Showers of purchases

62. German "genuine"

63. 80-Across's field

DOWN

1. Loc. of some devils

2. Tommie ___, 1966 A.L. Rookie of the Year

3. Used bookstore containers

4. Fill in the blank with this word: "___ Beach (D-Day site)"

5. Paint base

6. Some global treaty subjects, informally

7. Talk show host Lake

8. One who hears "You've got mail"

9. Winter falls

10. Camera operator's org.

11. English monarch who shared the throne

12. Have ___ in mind

13. "Time in a Bottle" singer

14. AM/FM device

18. Sack starter

22. The English translation for the french word: soma

24. Fill in the blank with this word: "___ Valley Conference"

25. You can't have your gateau (this) & eat it too

26. Realm of Otto I: Abbr.

27. Mancinelli's "___ e Leandro"

28. French key

29. The turkey tastes great!'

30. Ruler divs.

31. You might take stock in it: Abbr.

32. Tennyson's "immemorial ___"

33. Socks

34. Lb. or oz.

36. The English translation for the french word: poubelliser

37. Wreck-checking org.

38. Robert Louis Stevenson's "___ Triplex"

42. House finch

43. Miss Congeniality she's not

44. Kitchen gizmo

45. Words spelled out in currants on a Wonderland cake

46. "Hamlet" courtier

47. slender freshwater fishes of Eurasia and Africa resembling catfishes

48. With no help from the U.S., Simon turned for aid to this Caribbean nation that threw out the French in 1804

49. Fill in the blank with this word: "___ once"

50. Nightclub

53. Fill in the blank with this word: "Aid and ___"

54. When repeated, Mork's TV sign-off

55. Old Testament book

56. Zaire's Mobutu ___ Seko

58. Southeastern Conf. team

59. Warbler Sumac

PUZZLE 54

ACROSS

1. Wordsmith's ref.

5. Fill in the blank with this word: "___ Award (N.H.L.'s Coach of the Year trophy)"

10. This is ___'

13. Peek-___

14. Words from un innamorato

15. Explorer John and others

17. What a man making a comeback may get back to

19. Grand Ole ___

20. Spanish aunts

21. Mogul

23. Work on cud, say

26. Takes off

28. Where the X-axis meets the Y-axis

29. North Carolina university

30. Fill in the blank with this word: "___ Aquarids (May meteor shower)"

33. Author who wrote "One half of the world cannot understand the pleasures of the other"

34. Like the earliest Olympic festivals

35. Suffix with ball or bass

36. ___

37. wake (the opposite of)

38. Expy.

39. Saison d'___

40. Obtuse

41. an Indian side dish of yogurt and chopped cucumbers and spices

42. They may be fluid: Abbr.

43. Where to find the Wienerwald: Abbr.

44. Naysayer

45. Poster stock

47. Chest pain

48. Drank heavily

50. Voting no

51. Put ___ words

52. Human hand characteristic

58. Asian goat

59. Screwy

60. Fill in the blank with this word: "___ button (Facebook icon)"

61. Yellow ___

62. Unavailable for appointments

63. O.T. book before Daniel

DOWN

1. Morse T

2. There: Lat.

3. Trigonometry abbr.

4. Aim, e.g.

5. When many duels were held

6. Univ. paper

7. Popular cable channel

8. Middle of this century

9. Flattery

10. Spring blooms

11. Last monarch of France

12. Well, fiddle-dee-dee! He was the fifth emperor of Rome

16. Prefix with fuel

18. Fill in the blank with this word: "___ wait"

22. Life of Pi' author ___ Martel

23. Truck driving competition

24. Unreal

25. With 62-Across, theme of this puzzle

26. "Eeny-meeny-miney-mo" activity

27. Well-groomed

31. Letterman's nightly list

32. What is the capital of this country - Turkey

34. Fill in the blank with this word: ""I Still See ___" ("Paint Your Wagon" tune)"

37. Depression phenomenon

38. The English translation for the french word: tangible

40. With 25-Down, be prudent

41. Oscar de la ___

44. Cheers' star

46. Where drachmas were once spent

48. Was taken in

49. Much may follow it

50. Japanese golfer Isao ___

53. Kung ___ chicken

54. Tax fig.

55. La ___ (Hollywood nickname)

56. Fill in the blank with this word: "___ out a win"

57. William Shatner's '___War'

PUZZLE 55

ACROSS

1. Fill in the blank with this word: ""___ for the poor""

5. Swords, after conversion

15. make high-pitched, whiney noises

16. Settings

17. Homer

19. Fill in the blank with this word: "1099-___ (tax form sent by a bank)"

20. Time ___ half

21. Roman historian ___ Cassius

22. Fill in the blank with this word: "___ Rancho (suburb of Albuquerque)"

23. Ka ___ (Hawaii's South Cape)

24. Took on

28. U.S. Open champ, 1985-87

30. When clocks are set ahead: Abbr.

32. Wear down

33. Period when dinosaurs became extinct

35. Stub ___

38. Worker in the TV biz

39. What happened when I spilled coffee?

42. Ming of the Houston Rockets

43. The English translation for the french word: omettre

44. TV actor Katz

45. Red indication on a clock radio

47. Zoo critter

49. Redeemable recyclables

52. Unscramble this word: dsesaie

55. Francis Poulenc's "Le ___ masqu

57. Fill in the blank with this word: "___ one-eighty"

58. Radical 1970s grp.

59. You can get one on the house: Abbr.

60. TV control: Abbr.

61. Peaks

66. Fill in the blank with this word: "___ Line (German/Polish border)"

67. Wildcats' org.

68. Viking competitions?

69. Millennium divs.

DOWN

1. Spinning

2. Singer Kazan

3. Comic strip character's book about butchers' cuts?

4. Fill in the blank with this word: "___-Pitch"

5. Legislative assemblies

6. Domineered, with "it"

7. Two times tetra-

8. Fill in the blank with this word: ""Do ___ Diddy Diddy" (1964 #1 hit)"

9. Racing vehicles

10. They played Ricky Nelson and Bobby Darin

11. Went to Wendy's, say

12. Oysters ___ season

13. U.S. trading partner, formerly

14. Wind dir.

18. San Pablo Bay city

24. The U.N.'s Kofi ___ Annan

25. "Don't mess with the Hurricanes!," e.g.?

26. Mahler's "Das Lied von der ___"

27. Wagner's "___ fliegende Hollander"

29. Rap's Dr. ___

31. Took responsibility for

34. Super Bowl XXXVI and XXXVIII MVP Brady

36. Fill in the blank with this word: "___ pardo (grizzly, in Granada)"

37. See 35-Across

39. Suspense novelist ___ Hoag

40. Yeats's land

41. Reactor overseer: Abbr.

42. Torah place marker

46. Salutation abbreviation

48. undiscerning (similar term)

50. Writer Peggy known for the phrase "a kinder, gentler nation"

51. Zesty toppings

53. Let ___

54. Morley of "60 Minutes"

56. Lady ___, Girl Guides founder

59. Its coat of arms features a horseman spearing a dragon

61. Whole lot

62. Prefix with meter

63. Pay-___-view

64. Fill in the blank with this word: "___ Farrow, Mrs. Sinatra #3"

65. Mandela's org.

PUZZLE 56

	1	2	3	4	5	6		7	8	9	10	11	12	

(grid)

ACROSS

1. Provides a seat for

7. The English translation for the french word: posada

13. "You sure said it!"

14. Author of the popular 'Goosebumps' books

16. Less subdued

17. Made of certain twigs

18. Ransom ___ Olds

19. Fasten firmly, in a way

21. Fill in the blank with this word: "Capitol-___ (music company)"

22. Fill in the blank with this word: "___ prius (trial court)"

24. Take no action regarding

25. Sgts., e.g.

26. Mexican silverwork center

28. Super ___ (video game name)

29. Fill in the blank with this word: ""...___ man put asunder""

30. Fourth spin: jackpot!

33. *Commercial for a private school?

34. Some army exercises

40. Fill in the blank with this word: ""___ little silhouetto of a man" ("Bohemian Rhapsody" lyric)"

41. Most miserable hour that ___ time saw': Lady Capulet

42. Fill in the blank with this word: "___ knot, used in rug weaving"

43. Snoozes

44. 'The Venice of the Middle East'

46. Fill in the blank with this word: ""Let's ___""

47. Horatio! ___ do forget myself': Hamlet

48. Felt

50. Writer Santha Rama ___

51. Small and weak

53. Exert some pull

55. The English translation for the french word: hausser

56. Sea of ___

57. Wipes out

58. Cleared

DOWN

1. And other women: Lat.

2. He became king at age 5

3. You may be asked to arrive 90 mins. prior to this

4. Wolfe coined the term "radical" this in a story on a party for the Black Panthers thrown by Leonard Bernstein

5. Asian weight units

6. Infant's dessert, maybe

7. Inquire about a union contract?

8. Pitcher Gregg ___

9. Takes root

10. Wore away

11. Major digital satellite service provider

12. Windflower

13. William Morris workers

15. This inventor who never attended college has a college named for him in Trenton, New Jersey

20. Subject of the 1997 best seller "Into Thin Air"

23. Periods in contrast to global warming

25. Requiring more support

27. Fill in the blank with this word: ""___-Year Day" ("The Pajama Game" tune)"

29. They test reasoning skills: Abbr.

31. Was on the bottom?

32. Where B'way is

34. Cassiterite, e.g.

35. Usually

36. Grumbler

37. The English translation for the french word: chorale

38. Intertwined

39. The English translation for the french word: salut

44. Wallop

45. ___ boy

48. Spanish aunts

49. Ici ___ (here and there, to Th

52. Wilson Dam org.

54. Sue Grafton's '___ for Ricochet'

PUZZLE 57

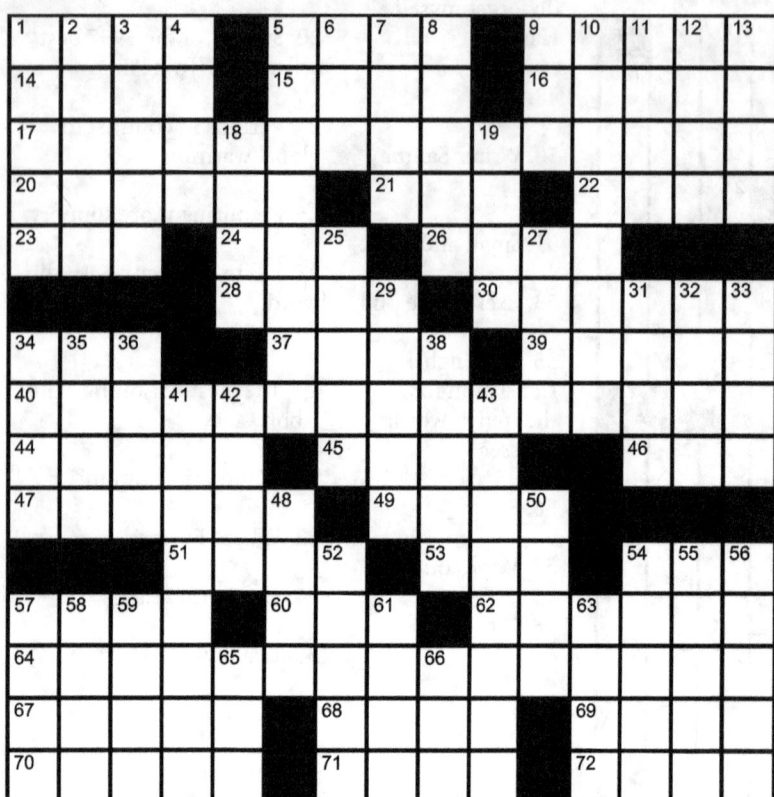

ACROSS

1. Tennis whiz

5. Tropical tuber

9. Play by a different ___ rules

14. Prix de ___ de Triomphe (annual Paris horse race)

15. Fill in the blank with this word: ""___ le roi!""

16. Where drachmas were once spent

17. Quote, part 2

20. The English translation for the french word: anorak

21. Fill in the blank with this word: "___ Park, home for the Pittsburgh Pirates"

22. Words with a nod

23. Whistler, at times

24. In 1986 U.S. Steel changed its name to this 3-letter one

26. Urge

28. Fill in the blank with this word: "___-Pei (dog)"

30. Fill in the blank with this word: ""Pretty ___""

34. Usher's offering

37. Fill in the blank with this word: "___-Day"

39. The "L" of A.F.L.-C.I.O.

40. Courage seeker in a 1939 film

44. Watchers

45. Tic-tac-toe choice

46. Volkswagen model

47. Value

49. What ___

51. Like llamas

53. The English translation for the french word: OMC

54. Fill in the blank with this word: "Doo-woppers ___ Na Na"

57. Suffragist Carrie Chapman ___

60. Fill in the blank with this word: ""Don't tell ___ can't ...!""

62. Yanks' org.

64. One who has practiced his hitting skills

67. Town council president, in Canada

68. Water barrier

69. Speed skater Apolo Anton ___

70. More desertlike

71. Katharine's role in "Adam's Rib"

72. Wilts

DOWN

1. ___ boy

2. Fill in the blank with this word: "Chili con ___"

3. Fill in the blank with this word: "___ Good Feelings"

4. Stereo syst. component

5. Place to plug a new book, maybe

6. Washboard ___

7. Woodworking tool

8. "Sesame Street" tune, with "The"

9. On the Road' narrator ___ Paradise

10. The best you can be, Freudian-style

11. Tugboat services

12. The English translation for the french word: orle

13. Tammy ___ of 1970s-'80s TV

18. Waters, informally

19. Old English bard

25. Popular antianxiety drug

27. Unite formally

29. Arrange into new lines

31. It's not ___ deal'

32. The English translation for the french word: suie

33. About

34. To ___ (unerringly)

35. John ___-Davies of the "Lord of the Rings" trilogy

36. Track ___

38. Play to ___ (tie)

41. Outside-the-box

42. Douay prophet

43. Fill in the blank with this word: "___.com (e-mail address for the Coast Guard)"

48. They say this lady will "coax the blues right out of the horn" & "charm the husk right off of the corn"

50. Seat of Allen County, Kansas

52. Canio's wife in "Pagliacci"

54. Fill in the blank with this word: ""I Am ... ___ Fierce," #1 Beyonc"

55. When doubled, a former National Zoo panda

56. B and O figures: Abbr.

57. They play in front of QBs

58. Answer to the riddle "Dressed in summer, naked in winter"

59. Six-foot vis-

61. Fill in the blank with this word: ""___ to please""

63. Those, to Tom

65. Fill in the blank with this word: "___ Tamid (synagogue lamp)"

66. Dog doc

PUZZLE 58

ACROSS

1. Mrs. Lincoln's maiden name

5. Year in John XVIII's papacy

9. Singer McEntire and others

14. Singer nicknamed 'The Jezebel of Jazz'

15. Gear teeth

16. Filled (with)

17. Way it's done

18. -

19. Stuffed ___ (kishke)

20. Bullies

23. Scott Joplin's "Maple Leaf ___"

24. In general

25. Revenuers

26. Young haddocks

29. Shakespeare's ___ of Salisbury

31. Newspaper's ___ page

32. Cleopatra's last request?

37. Where to see a mummy: Abbr.

38. Comment after a difficult decision

40. Fill in the blank with this word: ""Days of ___ Lives""

41. It's hard to understand

43. Woes of the world

44. Word before face or heart

45. Small bay

47. Vingt-___ (blackjack)

49. Patella

53. Fill in the blank with this word: "___ sgt. (police rank)"

54. Survivors' concerns

58. The English translation for the french word: sec

60. Writer LeShan and others

61. Ones in charge: Abbr.

62. Twisted humor

63. '60s protest / Skip, as a dance

64. Fill in the blank with this word: ""___ kleine Nachtmusik""

65. German indefinite article

66. Fill in the blank with this word: "Eye ___"

67. ___ le roi!' (French Revolution cry)

DOWN

1. Fill in the blank with this word: "___ of the Unknowns"

2. Whiff

3. Woodworking groove

4. Natural tint source

5. Radio couple at 79 Wistful Vista

6. relating to the palm of the hand or the sole of the foot

7. Marvin Gaye's "Can ___ Witness?"

8. Fill in the blank with this word: "Dan ___, former N.B.A. star and coach"

9. Some mail designations: Abbr.

10. We'll teach you to drink deep ___ you depart': Hamlet

11. Air show maneuver

12. Sportscaster Rashad

13. Some wild parties

21. Underwear initials

22. Like non-oyster months

25. To and ___

26. Fill in the blank with this word: ""___ nerve!""

27. The "brains" of 58-Down

28. Unscramble this word: slntoeuroi

30. Where the outboard motor goes

32. Fill in the blank with this word: ""___ was saying Ö""

33. Year that Dionysius of Halicarnassus is believed to have died

34. Fill in the blank with this word: "___ green"

35. Uncooperative one

36. Fill in the blank with this word: ""The even mead, that ___ brought sweetly forth ...": "Henry V""

38. What "y" might become

39. Wilderness walks

42. Whole lot

43. Spears

45. Scores 100 on a test

46. Kangaroo ___

47. 1983: "____ and the Cruisers"

48. Fill in the blank with this word: "___ Clark who sang "Poor, Poor Pitiful Me""

50. Weapons check, in brief

51. Fill in the blank with this word: ""Embraced by the Light" author Betty J. ___"

52. The Louvre's Salles des ___

54. Fill in the blank with this word: "___ Grand (supermarket brand)"

55. Marked, as a questionnaire box

56. About

57. Fixed at an acute angle

59. WSW's reverse

PUZZLE 59

ACROSS

1. Vanquished

5. Actress Samantha

10. Unmelodic sounds

14. Mythical king of the Huns

15. When one might have a late lunch

16. Bothers

17. Perfect plot

20. Persistence of memory concept

21. Plumbs the depths

22. Waters, informally

24. Part of an Asian capital's name

25. Fill in the blank with this word: ""Bali ___""

28. Texans' grp.

29. Either of two books of the Apocrypha: Abbr.

30. QB Detmer and others

33. About

35. Klinger player on "M*A*S*H"

37. Ready or not, here ___'

39. Hank Ketcham comic strip

42. Tony who led the N.L. in batting eight times

43. Unscramble this word: neam

44. Vardalos and Peeples

45. U.S. trading partner, formerly

46. The Missing Drink :

High _____ rose

48. Title for one on the way to sainthood: Abbr.

50. What's right in front of U

51. Life of Pi' author ___ Martel

52. CPR pros

54. Frog sounds

57. West Flanders resort port

61. Chick lit book #1 (1992)

65. Fill in the blank with this word: "___ temperature"

66. Worrisome food contamination

67. Rare book dealer's abbr.

68. Kipling's "___ we forget!"

69. Like some tattooed characters

70. I. M. and Mario

DOWN

1. Restrain

2. People: Prefix

3. Without ___ to stand on

4. Un gato grande

5. Winter headgear

6. Ensured: Abbr.

7. WorldCom competitor

8. Grass part

9. Places to sleep

10. Heart of France

11. Wing: Abbr.

12. It may rock you to sleep

13. Princes, e.g.

18. Tiny battery

19. a person devoted to refined sensuous enjoyment (especially good food and drink)

23. Fill in the blank with this word: ""Everybody Jam!" singer ___ John"

24. Make grief-stricken

25. Waffle

26. Fill in the blank with this word: ""___ there yet?""

27. Fill in the blank with this word: ""Back ___" (1974 Genesis song)"

30. Surface-___

31. Workout spots, for some

32. Fill in the blank with this word: ""What thou ___, write": Revelation"

34. The Admiral Benbow ___ ("Treasure Island" locale)

36. Line score letters

38. This is ___'

40. The English translation for the french word: incorporation

41. of or relating to or involved the practice of aiding the memory

47. Fill in the blank with this word: "Eve ___, "The Vagina Monologues" monologist"

49. Wreck-checking org.

51. Vocally bother

53. New York ___

54. Fill in the blank with this word: "___-Alt-Del"

55. Actress Madlyn

56. 1936 Olympics hero

58. Fill in the blank with this word: ""___ kleine Nachtmusik""

59. 40's theater director James

60. R.A.F. awards

62. Smallest NATO member by population

63. Voting "nay"

64. Muhammad ___

PUZZLE 60

ACROSS

1. Tiddlywink, e.g.

5. Subjective pieces

10. Sodium hydroxide, to chemists

14. San ___, Italy

15. Singer-songwriter Jones

16. Penny-___ (trivial)

17. Fill in the blank with this word: ""___ (So Far Away)" (1982 hit by A Flock of Seagulls)"

18. 1962 war epic, with "The"

20. Miami's ___ Bay

22. Wharves

23. Italian emporium ending

24. France's Cote d'___

26. Reaction from one who has a bone to pick?

29. Mythological hunter turned into a stag and killed by his own dogs

32. London lockups

33. Red as ___

34. Lana Del ___, singer with the 2014 #1 album 'Ultraviolence'

36. Landers and others

37. Longtime Yes drummer

38. The English translation for the french word: saisir

39. Rock's ___ Soundsystem

40. British runner Steve

41. Wake Island, for one

42. Tractor-drawn fall activity

44. Playwright Sean

45. Cub #21 of the 1990s-2000s

46. Like an N.B.A. team

47. Winningest southpaw in major-league history

50. Skateboard wheel material

54. Where to belt one down and belt one out

57. Massachusetts' ___ College

58. Fill in the blank with this word: ""Which Way ___?" (1977 film)"

59. Title for Sulu on "Star Trek": Abbr.

60. Walk back and forth

61. You can get one on the house: Abbr.

62. Std. on food labels

63. Squealed cries

DOWN

1. Wee bit

2. Yesterday, in Italy

3. Shirt sizes

4. Masks

5. Not unless

6. City ESE of Bombay

7. White-tailed eagle

8. The U.N.'s ___ Hammarskj

9. Fill in the blank with this word: "___-wolf"

10. Unscramble this word: tnuaer

11. Time ___ half

12. Buckwheat's affirmative

13. Attention-getters

19. Start of a weightlifting maneuver

21. Some airport data: Abbr.

24. Not worth ___

25. Beginning on: 2 wds.

26. Fill in the blank with this word: ""I've Got ___ in Kalamazoo""

27. Where you're likely to see dirty hands

28. "The Grapes of Wrath" star, 1940

29. Weaken

30. Two-tone treats

31. Writer Zora ___ Hurston

33. Tree-lined promenade

35. Every 12 mos

37. Rara ___

38. Light, one-seated carriage

40. Waste

41. Room to swing ___

43. A wishbone has one

44. Mario Puzo best seller

46. Whistle-blower's exposure

47. Whole alternative

48. Yesteryear

49. Fill in the blank with this word: "Cut ___"

50. Fill in the blank with this word: ""Deutschland ___ Alles""

51. Zoological wings

52. Small cut

53. Hydrocarbon suffixes

55. Nickname of 1954 home run leader Ted

56. Sue Grafton's '___ for Evidence'

PUZZLE 61

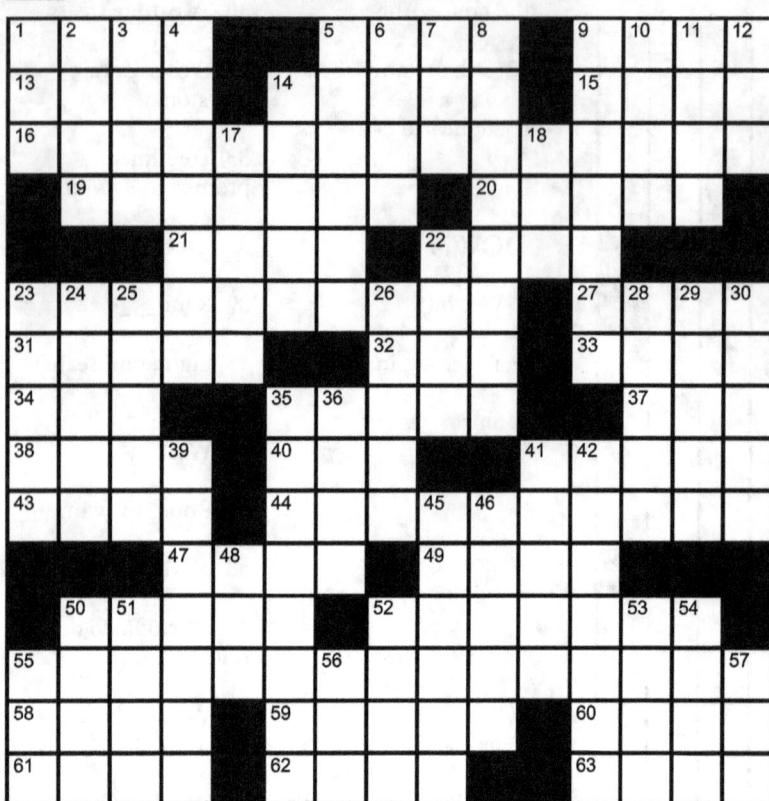

ACROSS

1. Retired, as a prof.

5. Poet Robert and others

9. Fill in the blank with this word: "___ dixit"

13. Voyager launcher

14. Fill in the blank with this word: "___ the heart of"

15. Scottish rejections

16. Quip, part 3

19. Its ruins are across the Tigris from Mosul

20. Fill in the blank with this word: "___ this world"

21. Critic, at times

22. Inventor Elias

23. Cruising, say

27. Fill in the blank with this word: "___ Vista"

31. Polo and others

32. Wharf workers' org.

33. Willy Wonka's creator

34. Way of the East

35. Literary alter ego

37. Fill in the blank with this word: ""Je te plumerai le ___" ("Alouette" lyric)"

38. Three men in ___'

40. Wanted-poster letters

41. Early statistical software

43. Novelist ___ Mae Brown

44. Flat turndown

47. Fill in the blank with this word: ""___ go!""

49. Jamaica's Ocho ___

50. Rice-___

52. Minds

55. "Not to my recollection"

58. Yawn

59. Optimistic

60. TV's 'How ___ Your Mother'

61. pretentious (similar term)

62. Fire

63. Wine casks

DOWN

1. Stationer's item: Abbr.

2. Spanish ___

3. They, in Italy

4. With 51-Across, wet-day wish

5. Nail-___ (tense situations)

6. Grateful Dead bassist Phil

7. Twist in the sky

8. Tablet

9. Unscramble this word: tidasen

10. Outdoor dining spot

11. When some Wimbledon matches are won

12. Yacht's dir.

14. Long-___

17. William Blake: "When the stars threw down their spears,/ And watered heaven with their _____"

18. What a violinist may take on stage, in two different senses

22. With gusto

23. Fill in the blank with this word: "___ of roses"

24. Hit so as to make collapse / Win over

25. Fill in the blank with this word: ""Indoors ___?""

26. Part of a drum kit

28. Jacob's father-in-law

29. What foxhounds try to catch

30. Rep. ___ Hastings of the House

intelligence committee

35. Decimal part of a logarithm

36. Old theaters once owned by Howard Hughes

39. pretentious or silly talk or writing

41. Hindu's loin cloth

42. This could raise a pitcher's 51-Across

45. The English translation for the french word: Ariane

46. having unsuitable feminine qualities

48. Treebeard, e.g.

50. Rent-___

51. Wholly absorbed

52. Wildcat with tufted ears

53. Movie whale

54. Fill in the blank with this word: "Explorer ___ Anders Hedin"

55. Supermarket with a red oval logo

56. Fill in the blank with this word: ""___ Wiedersehen""

57. Visitors to the Enterprise

PUZZLE 62

ACROSS

1. Sample a flavor

6. Swedish imports

11. What's right in front of U

14. Model railroad track measure

15. To whom a Muslim prays

16. Mount ___, active Philippine volcano

17. DEADLINE

20. Troll dolls, once

21. Sports org.

22. Slip on paper

23. Keto-___ tautomerism (organic chemistry topic)

25. Alcatraz inmate

27. Worker's demand

28. 54703

31. They dug his grave ___ where he lay': Sir Walter Scott

32. Yardsticks: Abbr.

33. The English translation for the french word: if commun

34. Fill in the blank with this word: ""The even mead, that ___ brought sweetly forth ...": "Henry V""

35. Torment

38. Launch ___

40. Fill in the blank with this word: ""___ U""

43. Victory sign

45. French painter Charles Le ___

48. Fill in the blank with this word: "China's Sun Yat-___"

49. "I, Robot" author

53. Prelate's title: Abbr.

55. The rich man in "Rich Man, Poor Man"

56. Zaragoza's river

57. Thin ___

59. Rock's ___ Speedwagon

61. The English translation for the french word: rut

62. Summit success

65. Star of "Youngblood," 1986

66. Seacrest's 'American Top 40' predecessor

67. Stevens of TV's "The Farmer's Daughter"

68. Fill in the blank with this word: ""___ help a lot!""

69. Support staff: Abbr.

70. Speed ___

DOWN

1. Shade of brown

2. Tropical lizards

3. A sponge may get this

4. Fill in the blank with this word: "Bronze ___"

5. Fill in the blank with this word: "Arctic ___"

6. Tubular food

7. Municipal pol.

8. Property recipients

9. "Roseanne" star

10. Fill in the blank with this word: "Children's writer ___ L. Smith"

11. More run-down

12. Other halves

13. The English translation for the french word: tourmenter

18. Rubaiyat' rhyme scheme

19. Mike Ovitz's former co.

24. Entertainment center at many a sports bar

26. It may finish second

29. W.W. II vessel: Abbr.

30. When tank warfare began: Abbr.

34. Weird

36. Tubes

37. Maintaining one's composure, say

39. USA alternative

40. Unscramble this word: seilair

41. Why bother?'

42. Not yet having gone before an M.P.A.A. board

44. Fill in the blank with this word: "___ minÈrale"

46. The English translation for the french word: ombrage

47. State of anarchy

49. One of the Trumps

50. They come with new computers

51. Yesterday, in the Yucat√°n

52. Select group?

54. We'll teach you to drink deep ___ you depart': Hamlet

58. Fill in the blank with this word: ""Take ___ a sign""

60. Of lyric poetry

63. This, to Th

64. Foot part

PUZZLE 63

ACROSS

1. Textbook market shorthand

5. "Hogan's Heroes" colonel

10. The rain in Spain

14. Riga resident

15. Using gold jewelry from the Israelites, he fashions the Golden Calf

16. Yesteryear

17. Man o' War

20. Bathtub sound

21. Picnic pests

22. Dole's running mate, 1996

24. Some Mercedes-Benzes

25. Fill in the blank with this word: ""Doo ___ (That Thing)" (#1 hit for Lauryn Hill)"

28. Novelist ___ Mae Brown

30. Unscramble this word: eatsmr

35. Fill in the blank with this word: ""___ never work!""

37. 1960 Updike novel

39. Where Ephesus was

40. Lafayette or Orleans

43. The English translation for the french word: fourmi

44. Spittoon sound

45. Indian percussion rhythm

46. Tiara

48. Some low-income housing, for short

50. O-___-O (brand of sponge)

51. Philosopher ___-tzu

53. The Runnin' Rebels of the N.C.A.A.

55. Tincture of opium

60. Fill in the blank with this word: ""___ on $45 a Day""

64. What a lessee often gets back less of

66. Handle: Fr.

67. Shorts material, in M

68. Greece's Mount ___

69. Youngsters

70. Fill in the blank with this word: "___ de la Frontera (town near C"

71. Fill in the blank

DOWN

1. Small toymakers: Var.

2. Quantum ___

3. www page creation tool

4. Fill in the blank with this word: ""___ a trip on a train..." (Benny Goodman lyric)"

5. of or relating to or characteristic of Kashmir or its people or culture

6. Fill in the blank with this word: ""Knots Landing" actress ___ Park Lincoln"

7. Some nest eggs

8. Prepares to shoot, as an arrow

9. Uses a hassock

10. Fill in the blank with this word: ""... poison'd with ___ of ale": Shakespeare"

11. Paris's ___ de Lyon

12. The Beatles' "Back in the ___"

13. Suffixes with sultan

18. Fill in the blank with this word: "End ___"

19. In Greek myth, she's the goddess of the hearth

23. Under-the-sink fitting

25. Tricked

26. Mexican Indian

with this word: "___ scale"

27. Feather in Juan's cap?

29. Some wedding guests

31. Loud noise

32. Univac's predecessor

33. Window alternative

34. Taj ___

36. Wasn't straight

38. World's smallest island nation

41. Of some monuments

42. Trekkers

47. The paper for this type of envelope was originally made from hemp of a certain Philippine city

49. Batter's bane

52. Plan 9 From ___ Space'

54. Souped-up engine sound

55. Would-be J.D.'s hurdle

56. Insurance giant

57. School in La Jolla: Abbr.

58. Union jack?

59. Ball V.I.P.'s

61. Fill in the blank with this word: "___ buco"

62. Word of contempt

63. Terminal info

65. Whittier war poem "Laus ___"

PUZZLE 64

ACROSS

1. The title role in "Boris Godunov" is for a singer in this vocal range

5. This verb comes from an Old English word for "tremble"; you might do it during a temblor

10. Finish this popular saying: "It is best to be on the safe_____."

14. The English translation for the french word: rite

15. Worrier's worry

16. Writer Bagnold

17. Reciprocal pronoun

19. Zipped through

20. Unscramble this word: suimtm

21. The English translation for the french word: ÈnormitÈ

23. Yesterday, so to speak

25. Wished undone

26. The English translation for the french word: inane

29. The English translation for the french word: opter

32. Unscramble this word: serds

35. The English translation for the french word: rat de bibliothÈque

38. To the ___ degree

39. Mysterious: Var.

40. Science

41. The English translation for the french word: stoa

42. Whiz

43. Toward that place; in that direction

45. Finish this popular saying: "The end justifies the_____."

47. Finish this popular saying: "Waste not want_____."

48. The last names of bestselling authors Arthur & Alex were both pronounced this way

49. This farewell word first appeared in an English text in Hemingway's "A Farewell to Arms"

51. Zipped

53. The English translation for the french word: astÈrisque

57. Sonny boy

61. Wolfe coined the term "radical" this in a story on a party for the Black Panthers thrown by Leonard Bernstein

62. An island in southeastern New York

64. The lovely locks of a Lipizzaner

65. Uneven

66. Old Icelandic literary work

67. Finish this popular saying: "Don't teach your Grandma to suck_____."

68. Shut out

69. The mood was

DOWN

one of this as the Moon broke from its orbit & hurtled toward Earth

1. Warner ___

2. The English translation for the french word: aÔnou

3. Watch part

4. Characteristic of or befitting a seaman

5. The English translation for the french word: quotitÈ

6. Last: Abbr.

7. Yearn (for)

8. Wail

9. Fill in the blank with this word: "Computer ___"

10. Tailoring machine

11. The English translation for the french word: incident

12. Unscramble this word: ited

13. Whirlpool

18. This Japanese term refers to second-generation Japanese-Americans, many of whom were interned during WWII

22. Fill in the blank with this word: ""How ___!""

24. A ribbed fabric used in clothing and upholstery

26. Steel girder

27. Great-___

28. Discharge bad feelings or tension through verbalization

30. Unscramble this word: oopht

31. The English translation for the french word: collant

33. Warehouse

34. Disreputable

36. Fill in the blank with this word: ""___ la la!""

37. The English translation for the french word: soude

41. Wrap in swaddling clothes

43. Rimsky-Korsakov's "The Tale of ___ Saltan"

44. These large birds can be found in mixed herds with Guanacos in South America

46. They may appear on a tree

50. unoiled (the opposite of)

52. Buy-one-get-one-free item?

53. #1 spot

54. Uneven hairdo

55. Upset

56. Turning point?

58. Woodworking groove

59. Fill in the blank with this word: "___-European"

60. Fill in the blank with this word: "___ cheese"

63. Troop grp.

PUZZLE 65

ACROSS

1. Fall mos.

5. Let ___

10. Fill in the blank with this word: ""___ first...""

14. U.S.S. Enterprise counselor

15. Us Weekly subject

16. Nearly

17. Wound application

18. Many a mall outlet

20. Employees in the sugar industry

22. Fill in the blank with this word: "Correo ___"

23. Mother ___

24. Little pooches

26. Hot film of 1947?

30. Supersharp knife

31. Rock's ___ Fighters

32. Old record label

36. Fill in the blank with this word: "Costa Rica's ___ Peninsula"

37. Unit of a legion

41. Whistler, at times

42. Verve

44. Lon ___ of Cambodia

45. Robert ___ Stevenson

47. Factory conduit

51. Paul Anka hit that made it to #19

54. Wild Indonesian bovine

55. Fill in the blank with this word: ""___ gut!" ("All right!"): Ger."

56. Alrighty then'

60. Playground shout

63. Spanish noblewoman ___ de Castro

64. Turgenev's birthplace

65. Top-quality

66. Fill in the blank with this word: "Decem ___ (Latin decade)"

67. Historic German admiral Maximilian von ___

68. The lion in "The Lion, the Witch and the Wardrobe"

69. Suspension

DOWN

1. Wish receiver

2. Writer ___ Stanley Gardner

3. Hairdo for Snooki of "Jersey Shore"

4. Theme of this puzzle

5. yield to another's wish or opinion

6. Newsman Jim

7. Spanish waves

8. Verdi's "___ giardin del bello"

9. Surveyor's dir.

10. Former Acura model

11. Italian flowers

12. Unscramble this word: rgeae

13. Baltimore team, in sportspeak

19. Garden ___

21. Fill in the blank with this word: ""___ Organum" (1620 Francis Bacon work)"

24. Janitorial tool

25. Silver coin of ancient Greece

26. Hair line

27. Fill in the blank with this word: "Auvers-sur-___, last home of Vincent van Gogh"

28. Transfer and messenger materials

29. Use www.irs.gov, say

33. Fill in the blank with this word: "___ to one's word"

34. Overlay with wood

or plaster

35. Fill in the blank with this word: ""The Bells ___ Mary's""

38. Landers and others

39. Fill in the blank with this word: "De ___ (again)"

40. Football Hall-of-Famer ___ Hirsch

43. "Drop City" novelist, 2003

46. Shortest book in the Old Testament

48. Wine: Prefix

49. Washington city, river or tribe

50. Small bridge limit, maybe

51. These: Sp.

52. Steep slope

53. Wistful exclamation

56. Trompe l'___

57. Tony-nominated choreographer White

58. Fill in the blank with this word: "___-Ration (dog food)"

59. "Scratch that!"

61. Work ___ sweat

62. They have Xings

PUZZLE 66

ACROSS

1. Biblical spy

6. Wellness org.

9. Lymphocyte found in marrow

14. Ropemaking fiber

15. Vitamin C source

16. Like oak leaves

17. Indian currency

20. Stock ticker inventor

21. Tulsa sch. named for an evangelist

22. Military branches: Abbr.

23. Stocking stuffers

25. Part of AMPAS

28. Fill in the blank with this word: ""Oh dear!" cried the waist, "___.""

34. Touch-tone 4

35. The Missing Drink : High _____ rose

36. Tease

37. Hollywood's Alan and Diane

40. Sue Grafton's "___ for Lawless"

42. Their ranks don't include DHs

43. Peter of "Lawrence of Arabia"

45. Violin cutouts

47. Op. ___ (footnote abbr.)

48. Medusa killer takes his agent to court?

52. Player of Det. Eames on "Law & Order: Criminal Intent"

53. Latin hymn "Dies ___"

54. Food preservative: Abbr.

57. Worldwide workers' grp.

59. You got that right!'

63. Disagree

67. Titled Turks

68. Brain and spinal cord: Abbr.

69. Fill in the blank with this word: "___ Lauder"

70. Whoops

71. 'My mama done ___ me ...'

72. Moves toward

DOWN

1. Unscramble this word: hsca

2. In ___ way

3. Peterson of 2003 news

4. Trick-taking game

5. Spook's break-in

6. Slangy turndown

7. Thought: Prefix

8. Fill in the blank with this word: "California's ___ Castle"

9. Slo-___ fuse

10. Old radio's ___ Stoopnagle

11. Ancient city with remains near Aleppo

12. Unscramble this word: afel

13. Ones in charge: Abbr.

18. Red as ___

19. Fit for a nobleman

24. Sew shut, as a falcon's eyes

26. Venerated image: Var.

27. Unscramble this word: lsle

28. Fill in the blank with this word: "Cole Porter's "___ Men""

29. King of old movies

30. Innocents

31. Great-___

32. Fill in the blank with this word: ""Let ___!" ("Go ahead!")"

33. Political cartoonist Thomas

34. Unappetizing fare

38. Two tablets, maybe

39. Slave to detail

41. Fill in the blank with this word: "Feng ___"

44. Jazzman Blake

46. Fin de ___ (remainder): Fr.

49. Vote for

50. "Bambi" author

51. Surrender a second time

54. Present time, for short

55. Fill in the blank with this word: "___ office"

56. Shade of blue

58. Wine: Prefix

60. Fill in the blank with this word: "___ time limit"

61. Yesterday, in the Yucatán

62. Ja and da

64. Bygone daily MTV series, informally

65. Worrying sound to a balloonist

66. Opium maker

PUZZLE 67

ACROSS

1. Fill in the blank with this word: "___ fide"

5. Windy City paper, with "The"

9. Johnny Cash's "___ the Line"

14. Spanish waves

15. Zu

16. Hamilton' actress ___ Elise Goldsberry

17. Jazz buff

18. Webzine

19. Oilman Kashoggi

20. Long after his Fauvist days, he designed a chapel at Vence, from stained glass down to the vestments

22. Like the pinky compared to the other fingers

24. Put ___ good word for

25. Ox, goat or sheep

26. Popular film Web site, briefly

29. Meadows of comedy

31. Wind dir.

34. Fussy dress

36. Sainted pope of A.D. 683

38. Odd sign at Men's Wearhouse?

41. Trotsky and Edel

42. Statistical grouping

43. Fill in the blank with this word: "___ Cayes, Haiti"

44. Sot's state

46. Original Dungeons & Dragons co.

47. Goddesses of the seasons

49. Lose __ whisker

51. Defeat by a stroke?

54. Potential Emmy nominees

58. Month between marzo and mayo

59. Fill in the blank with this word: "Faulkner's "Requiem for ___""

61. Davy Jones or any other Monkee

62. Krispy ___ doughnuts

63. Yanks' enemies

64. Yorkshire river

65. Fill in the blank with this word: "" ___

66. Put ___ on (limit)

67. Winnebago owner, for short

DOWN

1. "Golly!"

2. Greek goddess Athena ___

3. Computer type

4. Fill in the blank with this word: "___ art (text graphics)"

5. Writers' references

6. Finish this popular saying: "All roads lead to _____."

7. Sam Adams Rebel ___

8. Something often stubbed

9. Tehran native

10. Fill in the blank with this word: ""___ it!" (cry of accomplishment)"

11. Katherine _____ Porter

12. Where flocks feed

13. Fill in the blank with this word: "___ State"

21. Things gone awry

23. Wrigley Field flora

25. The English translation for the french word: brusque

26. Patsy Cline's "___ to Pieces"

27. 1957 hit for the Bobbettes

28. Texas ties

30. On the ___

31. successful (similar term)

32. Nobelist Bohr

33. Dentist's request

35. Swabs' grp.

36. Holmes and Hagman

37. Fill in the blank with this word: ""Humanum ___ errare""

39. Fill in the blank with this word: ""___, bro?""

40. Fidgety

44. Indian poet ___ Aurobindo

45. Soap opera actress Braun

47. *"If only!"

48. Young hooter

50. Fill in the blank with this word: "___ of roses"

51. Twelve ___ (Tara neighbor)

52. Shadow

53. Rare trick-taker

54. This instrument's name is from the Latin for "large war trumpet"

55. Tel ___, Israel

56. To laugh, to Lafayette

57. Slave to detail

60. Toshiba competitor

(top of page, continued clues)
Isn't So" (Hall & Oates hit)"

PUZZLE 68

ACROSS

1. Fill in the blank with this word: ""The ___ lama, he's a priest": Nash"

5. Used to express farewell, this French word means "I commend you to God"

10. Ming's 7'6" and Bryant's 6'6", e.g.: Abbr.

14. Tear down, in England

15. Flag features

16. Stratford-___-Avon

17. Old, deteriorated ship

19. Popular vacation locale

20. Fill in the blank with this word: "Disco ___ (character on "The Simpsons")"

21. Spy Aldrich ___

22. Venus or Mars

24. Son of, in Arabic names

25. Fill in the blank with this word: "___-Alpes (French department)"

29. Classic film duo

35. Seacrest's 'American Top 40' predecessor

36. Fill in the blank with this word: ""Laborare est ___" ("to work is to pray")"

37. Wilson Dam org.

38. Spanish aunts

39. Wind through Darwin, Minnesota or Cawker City, Kansas; they both claim the world's largest ball of this

40. Fill in the blank with this word: "___-fry"

41. Switch ups?

42. Ohio natives

43. Fill in the blank with this word: "Carry ___"

44. Prominent media member

47. Orange dwarfs

48. Stamps, say

49. Take in

50. Fill in the blank with this word: "___ time limit"

52. Addams portrayer, in film

55. Unscramble this word: lalw

58. From dawn till dusk

61. What writer's block may block

62. Planting

63. Wear down

64. Attention-getters

65. Stiff hairs

66. Unscramble this word: dere

DOWN

1. ___ Island (location near Portland, Maine)

2. Suffix with aqua

3. Fill in the blank with this word: ""Mi casa ___ casa""

4. Fill in the blank with this word: ""___ Me Call You Sweetheart""

5. Wellesley grad

6. Stout, freshwater fish

7. Varieties

8. Suffix with morph-

9. See 13-Down

10. loud confused noise from many sources

11. Family M.D.'s

12. Water tester

13. Winter weather, in Edinburgh

18. Child in a 1980s custody case

23. the crime of forcing a woman to submit to sexual intercourse against her will

24. Summer coolers

26. "For shame!"

27. Writer St. John ___

28. Web, at times

29. Clock sound

30. Montreal Expos legend Tim

31. How silverware is often sold

32. unsuccessful (similar term)

33. One with a dish towel

34. Medieval merchants' guild

39. Take unwanted steps?

40. Orch. section

42. Fill in the blank with this word: ""Just you wait, ___ 'iggins...""

43. Jim-dandy

45. Band with the 1975 #1 hit "One of These Nights"

46. Thick, creamy soup

50. Use a knife

51. Writer ___ St. Vincent Millay

52. Fill in the blank with this word: "___ de vivre"

53. Words often before a colon

54. Teen-___

55. Writer Fleming

56. Where Schwarzenegger was born: Abbr.

57. W.W. II vessel: Abbr.

59. Star of "Youngblood," 1986

60. These 3 letters refer to a company's liability, or an old Ford

PUZZLE 69

ACROSS

1. "Happy Days" fellow

5. Grp. organizing '60s sit-ins

9. When you "make" this, you go with speed

14. Swiss artist Paul ___

15. Unscramble this word: loto

16. Older couple's home, often

17. Like a dream

20. Wish granter

21. South of Brazil?

22. Without ___ (unprotected)

23. Latvia's capital

26. Trick

28. Cheap chat

34. They, in S

35. Fill in the blank with this word: "Carolina ___"

36. Scottish slopes

38. Muhammad ___

39. Weather map lines

42. Fill in the blank with this word: ""There is no ___ team""

43. Role-playing game, briefly

45. Three ___ match

46. Voting no

47. Merits at least a 20% tip?

51. Gas: Prefix

52. Yankee or Angel, for short

53. Dugout shelter

56. War on Poverty agcy.

58. Over here...'

62. Amnesiac's vague recollection of having a hobby?

66. Blakley of 'Nashville'

67. Russian ruler: Var.

68. Fill in the blank with this word: ""The ___ lama, he's a priest": Nash"

69. Wishes one can get on a PC?

70. Insurance giant

71. Fill in the blank with this word: ""___ #1!""

DOWN

1. Business school subj.

2. Golden ___ (century plant)

3. Working stiff

4. Long flights

5. Pou ___ (vantage point)

6. Zone, so to speak

7. Tent furniture

8. Seventh heaven

9. Fill in the blank with this word: ""___ Haw""

10. Aardvark

11. Muralist JosÈ MarÌa ___

12. Encouraging sign

13. Vous ___ ici'

18. Trompe l'___

19. Yves Klein found this heavenly color a symbol of pure spirit & made works that were just a field of it

24. Transcript stats

25. Words of discovery

27. Highly toxic pollutants

28. Fill in the blank with this word: "___ poisoning"

29. Stewpots

30. Queeg's command

31. What fun!'

32. Psychiatrist/author R. D. ___

33. Scottish seaport

37. ___ le roi!' (French Revolution cry)

39. Fill in the blank with this word: ""If only ___ known ...""

40. Wild Indonesian bovine

41. Outfielder Mondesi

44. Sink-side rack

46. Place for rolls

48. Switch suffix

49. Time out?

50. Weather info: Abbr.

53. Yorkshire river

54. Varsity QB, e.g.

55. Femme fatale in "The Carpetbaggers"

57. Trickle

59. Fill in the blank with this word: "___ function"

60. One succumbing to 6-Down

61. "Sleeping" sensation

63. Turner of TV channels

64. Finish this popular saying: "Time and tide wait for no_____."

65. Victorian ___

PUZZLE 70

ACROSS

1. Working in a mess

5. Fill in the blank with this word: "Darin and Dee's "___ Man Answers""

8. Stiller and ___ (comedy duo)

13. Strip joints?

15. Fill in the blank with this word: "___ juris"

16. The E of Euler's formula V + F - E = 2

17. Course for course preparers

18. Weight abbr.

19. Passing remarks?

20. Astronaut's sign-off?

23. Beginner

24. Fill in the blank with this word: "___ sch."

25. See 35-Across

27. Flash ___ (faddish assembly)

30. William Saroyan's "My Name Is ___"

32. ___ Robbins, co-lyricist of the #1 "Rocky" theme song "Gonna Fly Now"

33. Soup brand

35. U.K. carrier, once

37. Fill in the blank with this word: "___ Polo of "Meet the Fockers""

41. 1961 #1 hit for Bobby Lewis

44. Fill in the blank with this word: "Bouquet ___"

45. Work of prose or poetry

46. The out crowd

47. Morse T

49. Viennese-born composer ___ von Reznicek

51. Fill in the blank with this word: ""Is it soup ___?""

52. Schoolmarm's impartation

56. Answer to the riddle "Dressed in summer, naked in winter"

58. Want-ad letters

59. Novelist-critic dances

64. Numbers game

66. Pres. Hoover's dog King ___

67. Unscramble this word: astrt

68. Require salting, maybe

69. Fill in the blank with this word: "Dryden's "___ for Love""

70. *"If only!"

71. Whom Jimmy once courted off court

72. Fill in the blank with this word: "At a low ___"

73. D.O.E. part: Abbr.

DOWN

1. Fill in the blank with this word: "___ Rios, Jamaica"

2. Movie whale

3. Linens for a large bed

4. French versifier

5. Home of the Shah Faisal Mosque

6. Clothing company since 1992

7. Oise tributary

8. Quaint exclamation

9. Teacher's deg.

10. I Kissed ___' (Katy Perry hit)

11. Tighten, as laces

12. Unscramble this word: aetss

14. Hanks's "Bosom Buddies" co-star

21. Celtic sea god

22. Perform as one

26. L'___-deux-guerres (French era)

27. Business school subj.

28. Fill in the blank with this word: "___ off (switch choice)"

29. Cap'n's mate

31. There's a Cafe du this, "the world", in New Orleans

34. Where to sign a credit card, e.g.

36. Without wavering

38. Fill in the blank with this word: ""Just you wait, ___ 'iggins...""

39. Finish this popular saying: "If wishes were horses, beggars would___."

40. Research facility: Abbr.

42. This was Indira Gandhi's maiden name (her father was India's first P.M.)

43. Turn loose

48. Fill in the blank with this word: "___ Fables"

50. Fill in the blank with this word: ""Just Another Girl on the ___" (1993 drama)"

52. Goal-oriented

53. The English translation for the french word: hooch

54. Spot remover?

55. Relatively cool red giant

57. Excite

60. Fill in the blank with this word: ""___ Lang Syne""

61. Yards advanced

62. It's you ___'

63. Fill in the blank with this word: "___-fry"

65. Honey eater of New Zealand

PUZZLE 71

ACROSS

1. Massachusetts Ave. bldg., in D.C.

4. London line

9. Thumbs through

14. Michigan's ___ Canals

15. Take a piece from

16. When you pull into Guadeloupe, you may have to change your dollars into this official currency

17. Commuter aircraft, maybe

19. Fill in the blank with this word: "___ Kristen of "Ryan's Hope""

20. Visigoth king who sacked Rome

21. Fill in the blank with this word: ""It ___

Necessarily So""

23. Probe persistently

24. Tu-144 and others

26. Fill in the blank with this word: "E pluribus ___"

30. Fill in the blank with this word: "___ place"

31. Where stranded canoeists get together?

33. What a welcome sight relieves

35. Fill in the blank with this word: "1099-___ (tax form sent by a bank)"

36. Spice in Christmas cookies

39. Rates of return

41. Uganda's ___ Amin

42. Closest

46. Stuffed with ham and cheese and then saut

48. Puccini soprano

52. Granny ___

53. Say "%@&#!"

54. Fastenable, as labels

55. ___ Theme,' tune from 'Star Wars: The Force Awakens'

57. Title bandit in a Verdi opera

58. Fill in the blank with this word: "___ Gonz"

61. Oafish

64. Unscramble this word: vtisi

65. Fill in the blank with this word: "Botticelli's "The Birth of ___""

66. Suffix with hotel

67. Trial figure

68. Ltr. accompaniers

69. Vitamin stat.

DOWN

1. Tabasco, por ejemplo

2. Fill in the blank with this word: "___ Rouge"

3. young leaves eaten in salads or cooked

4. Ring tossed at pegs

5. Sexist, say

6. Tin ___

7. Tail: Prefix

8. Understanding

9. Fill in the blank with this word: "1993 best seller "___-Language""

10. Civilization, to Hesse

11. Wrath

12. Fill in the blank with this word: "Dieu et ___ droit (motto of England)"

13. Wind dir.

18. Rum, vodka and orange juice drink

22. Words of discovery

24. Harpoon

25. Sault ___ Marie

27. Fill in the blank with this word: "Actor ___ Patrick Harris"

28. Flip

29. Trading places: Abbr.

31. Pressure

32. Some CBS forensic spinoffs

34. Wine: Prefix

36. Small cut

37. Tempura ___ (Japanese dish)

38. Novice: Var.

40. Her theme song was "Love Me or Leave Me"

43. Insulting

44. Went off

45. Fill in the blank with this word: "___ amis"

47. Public transportation to New York's Yankee Stadium

49. Salty gulp

50. Bamboozled

51. "Peer Gynt" dancer

54. The English translation for the french word: tresse

56. Within: Prefix

57. Month preceding Rosh Hashanah

58. Tiny energy units, for short

59. Kiddie ___

60. Vapour trail?

62. Pitcher Robb ___

63. S.A.S.E., e.g.

PUZZLE 72

ACROSS

1. Fill in the blank with this word: "___ corn"

6. TV's Houston and Dillon

11. Give nothing to

14. The second part missing in the author's name ___ Vargas ___

15. Fill in the blank with this word: ""Blame It ___" (Michael Caine film)"

16. Book before Zephaniah: Abbr.

17. Old comic actress ___ Janis

18. 1973 Jim Croce album

20. Rescuee's declaration

22. Something left of center?

23. Fill in the blank with this word: "___ off (switch choice)"

24. Small egg

26. Get rid of

27. Nothing special

32. Subject, in Spain

33. Talking-___ (scoldings)

34. Where Noah landed

39. Forest denizens

41. Prepare mentally

42. Gems, precious metals, etc., in Spain

43. WSW's reverse

44. The English translation for the french word: rite

45. Trusted

48. Flag features

52. Canine shelter

53. Detestation

54. Tokyo, once

55. Useless

60. Platinum-selling 10,000 Maniacs album of the 80's

63. Wrinkle with age

64. Music category

65. Threepeater's threepeat

66. Fill in the blank with this word: ""___ could have told you that!""

67. U.S.N.A. grad

68. Fill in the blank with this word: "___ cake"

69. Tried to keep one's

DOWN

1. Fill in the blank with this word: "___ sch."

2. Unite formally

3. Fill in the blank with this word: "___ Hodesh"

4. Toddler's attire

5. North Atlantic's ___ Islands

6. Kenyan president Daniel arap ___

7. Orioles owner Peter

8. Saint-___, France

9. South American monkey

10. Fill in the blank with this word: ""___ wise guy, eh?""

11. Fill in the blank with this word: ""Come back, ___" (western line)"

12. "Let's go, Pedro!"

13. Movie critic Roger

19. Yuletide beverage

21. People stand for this

25. Some hairstyles

26. Fill in the blank with this word: ""Just you wait, ___ 'iggins...""

27. Periodic table no.

28. "My stars!"

29. Young troublemakers

30. Spring time in Lisbon

31. Dons effortlessly, as

seat

35. Top: Prefix

36. Nose: Prefix

37. On ___ (proceeding independently)

38. Formal hat, informally

40. De ___ (superfluous)

41. Tire, at the Michelin plant

43. Praise to the heavens

46. Ripley's love

47. Bit of a muscle car's muscle

48. Vouvray wines come from this valley noted for its chateaux

49. Oilman Kashoggi

50. Trunk attachments

51. Fill in the blank with this word: ""___ bien""

54. Italian emporium ending

56. The English translation for the french word: ruche

57. Weizman of Israel

58. Watercolorist ___ Liu

59. Vladimir Nabokov novel

61. Lumberjack's tool

62. Most miserable hour that ___ time saw': Lady Capulet

footwear

PUZZLE 73

ACROSS

1. "High Hopes" lyricist

5. P. C. Wren novel "Beau ___"

10. Architect of Spain

14. Tropical fever

15. Fill in the blank with this word: ""___ to you""

16. Fill in the blank with this word: ""A Letter for ___" (Hume Cronyn film)"

17. Song standard from Broadway's "Jubilee," 1935

20. Secure online protocol

21. Tommie ___, 1966 A.L. Rookie of the Year

22. Suitcase convenience

23. Environmental sci.

24. Role in "Son of Frankenstein"

25. Silence

28. Hides

30. Reply facilitator: Abbr.

33. Very, to Verdi

34. Ursine : bear :: pithecan : ___

35. Usually toasted sandwiches, for short

36. Very close, in a way

40. Unscramble this word: iphs

41. Sports org. with the Calder Cup

42. Trailer for farm animals?

43. Sot's sound

44. Tabby talk: Var.

46. Sales slips: Abbr.

47. "Now you're talking!"

48. Welsh symbol

50. Role for Ann Sothern in ten films

53. Where chamois and snow leopards live: Abbr.

54. Youth org. since 1910

57. Two forms of ID for Britney Spears?

60. Sinead O'Connor album 'Am ___ Your Girl?'

61. Fill in the blank with this word: ""Hello ___" (Todd Rundgren hit)"

62. Sounds made by 36-Across

63. Watch readouts, for short

64. Lack of musical talent

65. Roman ___

DOWN

1. Yellow fleet

2. Fill in the blank with this word: "Dark ___"

3. Nicole's co-star in 'Australia'

4. Verdi's "___ giardin del bello"

5. Very beginning

6. Waters and others

7. Go on a vacation tour

8. Hot ___

9. Ornate centerpieces

10. Made off with

11. Davy Jones or any other Monkee

12. Twining stem

13. Will who played Grandpa Walton

18. Tortilla triangles

19. Sticky substances

23. Thwart in court

24. Cry of pain

25. #1 song

26. Fill in the blank with this word: "___ Minh"

27. First saint canonized by a pope

29. Fill in the blank with this word: "Artist Frida ___"

30. ___ wake (the opposite of)

31. Wiped out, slangily

32. These: Sp.

35. Unscramble this word: kbcir

37. Came to port

38. Fill in the blank with this word: ""___ she blows!""

39. Mother ___

44. Year of Philip I's birth

45. Fill in the blank with this word: "___ Thule, distant unknown land"

47. Support staff: Abbr.

49. Fill in the blank with this word: "___ into holy matrimony"

50. Work hard

51. Nary ___

52. This device from Apple can put 10,000 songs in a music fan's pocket

53. Warehouse contents: Abbr.

54. Star of "Charles in Charge"

55. Tastes

56. Handle: Fr.

58. Fill in the blank with this word: ""Am ___ believe Ö?""

59. Ob-___ (med. specialty)

PUZZLE 74

ACROSS

1. Fill in the blank with this word: "Diplomat Boutros Boutros-___"

6. Madrid month

10. Social slight

14. Use the spade again

15. Sharon of 'Dreamgirls'

16. Unexciting

17. From the start

18. Onetime Robin Williams co-star

20. Temporary skylight?

22. Fill in the blank with this word: "___ Morris, signature on the Declaration of Independence"

23. Suffix with morph-

24. Purveyor of nonstick cookware

26. Idea destined to fail

28. Funnies flier

32. Wonderful' juice brand

33. Turkeys

34. Feeling, Italian-style

38. Yemen's Gulf of ___

40. Some H.S. exams

43. Unscramble this word: ache

44. Fill in the blank with this word: ""What should I ___?""

46. Dragsters' org.

48. Reply facilitator: Abbr.

49. T. S. Eliot work

53. Son of Mary Stuart

56. Shakespeare's ___ of Salisbury

57. Shade of gray

58. Jazz saxophonist/flutist Frank

60. Reach, as great heights

64. It may get you slapped

67. Fill in the blank with this word: "Edgar Bergen's Mortimer ___"

68. Prefix with -stat

69. Go-aheads

70. Exuded

71. Fill in the blank with this word: "___ Gailey of "Miracle on 34th Street""

72. Way from Syracuse, N.Y., to Harrisburg, Pa.

73. They test reasoning skills: Abbr.

DOWN

1. Weight of a paper clip, roughly

2. Nectar-pouring goddess

3. Poke-___!' (kids' book series)

4. Much of "The Ed Sullivan Show"

5. Slip acknowledgment

6. Grp. with the 1973 gold album "Brain Salad Surgery"

7. Fill in the blank with this word: "___ East"

8. Irish runner Coghlan

9. Inner circle member

10. Yemen-to-Zimbabwe dir.

11. What 'check' could mean

12. 1972 #2 hit for Bill Withers

13. Uninvited cornfield guest

19. B and O figures: Abbr.

21. Incurred, as charges

25. Watch things, for short

27. Snick-or-___

28. W.W. I plane

29. Fill in the blank with this word: "___ list"

30. City with a Volkswagen plant

31. Yoga posture

35. Voyager launcher

36. The English translation for the french word: scannÈriser

37. Expressed surprise

39. Written reminder

41. Fill in the blank with this word: ""Just hear ___ sleigh bells jingling "

42. Spanish counterparts of mlles.

45. A list of the A-list

47. The English translation for the french word: aÈrosol

50. Plane, e.g.

51. Most sensible

52. South American plains

53. Regions

54. Shower

55. One of the three original Muses

59. Go on a vacation tour

61. Yasmina ___, two-time Tony-winning playwright

62. Waste allowance of old

63. Fill in the blank with this word: "Even ___"

65. Turf

66. Wind dir.

PUZZLE 75

ACROSS

1. With 29-Down, central role on "Knots Landing"
5. La Scala cheer
10. Wall Street whizzes
14. Singer Brickell
15. The English translation for the french word: taudis
16. Town NNE of Santa Fe
17. What wakes people up in Washington?
19. Vaulted space
20. Slugfest
21. What to call an archbishop
23. Son of, in Arabic names
25. The ruler of Qatar is known by this 4-letter title
26. Possible item in a window box
30. Fill in the blank with this word: ""___ Live," 1992 multiplatinum album"
32. Three-way joint
35. "Doesn't that beat all?!"
37. The English translation for the french word: mÈgoter
39. Italian emporium ending
40. Sluggish water
42. Certain plaintiff, at law
43. Really neat
45. End of the graffiti
47. Fill in the blank with this word: "___-noir (modern film genre)"
48. School ___
50. Fill in the blank with this word: ""___ Wedding," Alan Alda film"
51. Unscramble this word: liar
53. The English translation for the french word: …os
54. Animated series starring Jon Lovitz
58. Staffordshire stench
63. Salinger's 'For ___ - With Love and Squalor'
64. Like some shoes
66. Waterfall sound
67. Go ___ (turn in)
68. Painter Mondrian
69. Landers and others
70. Ukrainian port, to natives
71. Series of legis. meetings

DOWN

1. 1969 Beatles hit
2. Sixth Jewish month
3. Fill in the blank with this word: "Costa ___"
4. That hurts!'
5. Food preservative: Abbr.
6. Fill in the blank with this word: "___ Smith, first female jockey to win a major race"
7. Onetime Chevy subcompact
8. Soft palate
9. Attorney Gloria
10. Video game pioneer
11. Criticize creative types?
12. Winter pear
13. Go on a vacation tour
18. CCCLI tripled
22. Yukons, e.g.
24. The English translation for the french word: bulbe
26. Oscar winner Edmund of "Miracle on 34th Street"
27. Weird
28. Like the 1974 rope-a-dope fight
29. About 40 degrees, for N.Y.C.
30. Deep cavity
31. Theme of this puzzle
33. What might do a foul tip?
34. Sporting blades
36. Fill in the blank with this word: "Dragon's ___ (early video game)"
38. Rock's Cobain
41. Central knob of a shield
44. the earth from his flesh
46. Turhan Bey, as this "Fabled" author, spent "A Night in Paradise" with Merle Oberon but without a "Moral"
49. Former San Francisco Mayor Joseph
52. They provide excellent service
53. Repeat calls?
54. Trillion: Prefix
55. Personal and direct
56. Well-___
57. The Beatles' "Let ___"
59. Passbook amts.
60. Fill in the blank with this word: "___ Cologne (skunk of old cartoons)"
61. Turnarounds, slangily
62. Proofs of purchase: Abbr.
65. Harem room

PUZZLE 76

```
 1  2  3  4  5  6  7  8  9     10 11 12 13 14
15                          16
17                          18
19          20          21     22
23       24 25    26           27
      28    29       30        31
   32 33             34        35
36                37 38 39
40             41
42          43 44 45    46
47          48       49    50    51 52 53
54          55          56       57
58       59          60    61 62
63                64
65                66
```

ACROSS

1. With 31-Down, feature of this puzzle's theme entries

10. Modern viewing options, for short

15. Excoriates

16. United States biochemist (born in Spain) who studied the biological synthesis of nucleic acids (born in 1905)

17. Overexposure or redeye?

18. Reptilian, in a way

19. Fill in the blank with this word: "___ Solo of 'Star Wars'"

20. Wrap

22. Y. A. Tittle scores

23. Certain constrictor

26. Record listing

27. Two qtrs.

28. Winnie-the-Pooh's Hundred ___ Wood

30. French possessive

31. Yo te ___'

32. French painter (1684-1721)

35. Fill in the blank with this word: "Dona ___, 1978 Sonia Braga role"

36. NETWORK

40. W.C.'s

41. See 35-Across

42. Wore away

43. Don Knotts's alma mater: Abbr.

46. Truth or ___ (slumber party game)

47. Soldier

48. When you "make" this, you go with speed

50. Spanish direction

54. Ore-___ (frozen food brand)

55. Start of a MacArthur quote

57. Saccharin discoverer ___ Remsen

58. You have too much of this in your blood if you have diabetes mellitus

60. They're often found in parentheses

63. Two-time NOW president Eleanor

64. Word on a Sharpie

65. Vaulted areas

66. Many bloggers

DOWN

1. Fill in the blank with this word: "___, beta, gamma ..."

2. Prostitute who protected Israelite spies, in Joshua

3. Fill in the blank with this word: "___-Lodge"

4. Wormer, say

5. Paul Anka's '___ Beso'

6. Jamaica's Ocho ___

7. Vegan morsels

8. Stock phrase

9. Went nuts

10. Swindlers, slangily

11. Year in St. Sergius's papacy

12. False!'

13. Harry Potter villain

14. Continues

21. Weeder's tool

24. Robbins and others

25. Fill in the blank with this word: "Entr'___"

29. Whistler, at times

32. Classic Bette Davis line from "Beyond the Forest"

33. Valedictorians have them

34. What a lei person might pick?

35. U.S. Army training center in Va.

36. Fill in the blank with this word: ""___ Explains It All" (cable series)"

37. "The Lonely Goatherd" performers, e.g.

38. Fill in the blank with this word: ""An' singin there, an' dancin here, / Wi' great and ___": Burns"

39. Old magazine ___ Digest

43. Reels

44. Fill in the blank with this word: "___ deferens"

45. Horseshoe-like figure

49. Shipping weights

51. Teams

52. You have to figure that at some time in his life Prince Charles has made it to the city of Stoke-on-this

53. Bridge seats

56. Fill in the blank with this word: "Dalai ___"

59. Popular cable channel

61. Small, sandy island

62. Maritime CIA

PUZZLE 77

ACROSS

1. Sombreroed cowboy

7. Was appealing

11. XXX counterpart

14. With 33-Across, confirmation, e.g.

15. the introductory section of a story

16. a group of African language in the Niger-Congo group spoken from the Ivory Coast east to Nigeria

17. Odorless gas

18. Sore to the max

20. Fill in the blank with this word: ""Doesn't this strike you ___?""

21. Hoped-for result of a merger

22. Folk song that was a 1958 #1 hit

24. Had too much, briefly

27. Hydrocarbon suffixes

28. Swiss stream

29. Compound used to stabilize perfumes

31. Trading places: Abbr.

32. "Holy cow!"

33. "I must submit to an epitaph graven by a fool" penner

38. Tropical flea

39. Where Loews is "L"

40. Virginia, once

41. Uno, due, ___

42. Fill in the blank with this word: ""___, Pagliaccio" (aria)"

46. Short person?

47. Black and white #5

49. 1982 Grammy-winning singer for "Gershwin Live!"

52. Judaism : kosher :: Islam : ___

53. Small dosage amount

55. Housekeeping

56. Range part: Abbr.

57. Sexist, say

58. TV Guide's Pennsylvania headquarters

59. Stock units: Abbr.

60. Unscramble this word: epos

61. Sense

DOWN

1. Unscramble this word: rteeca

2. Tries to pick up

3. Where telecommuters work

4. Checks, as checks

5. Sonata section

6. Just recently

7. Rams' gate?

8. Tv Sidekicks : Squiggy

9. Fill in the blank with this word: "Competitive ___"

10. Writer Earl ___ Biggers

11. Alrighty then'

12. Cries of pain

13. Kellogg's Cracklin' ___ Bran

19. Outfielder's shout

21. Ticket

23. Shootin' Annie

25. D.O.E. part: Abbr.

26. Old African rulers

29. Purse items

30. Talk of the Gaelic

31. Ancient Greek sculptor of athletes

33. Shoe company founded in Denmark

34. Red Sea vessel

35. Frequent collaborator with Miles Davis

36. Town centers in old Greece

37. Ancient Greek sculptor famous for his athletes in bronze

42. Site of 1990s genocide

43. Honolulu's ___ Palace

44. Tolkien's Smaug, for one

45. Goof-offs

47. Veal cuts

48. The English translation for the french word: tonte

50. Fill in the blank with this word: ""___ and away!""

51. Fill in the blank with this word: ""This will ___ further""

53. They often accompany logos: Abbr.

54. Ordinal suffix

55. 1959 #1 hit for the Fleetwoods

PUZZLE 78

ACROSS

1. Greeter at the door

11. Songwriter Novello

15. Energize

16. Fill in the blank with this word: ""___ #1!""

17. There are six of these in the middle of 17- and 56-Across and 11- and 25-Down

18. Yard sale tag

19. Something to chew on

20. That ole boy's

21. Its slogan was once "More bounce to the ounce"

22. You'll want to munch on petha & gazak, signature sweets of this Taj Mahal city

24. Nautical direction

27. Pizzeria ___ (fast-food chain)

28. "Quiet!"

32. Sports org. that publishes DEUCE magazine

33. Whse. unit

34. TV remote, e.g.

36. Fill in the blank with this word: "___ in Charlie"

37. It may be pulled

41. Mother ___

42. Deep fissure

43. Command level: Abbr.

44. Kellogg's Cracklin' ___ Bran

45. Your line of fate is quite deep, indicating success investing with tech stocks, like Adobe & Oracle, on this exchange

48. Fill in the blank with this word: "___ Aquarids (May meteor shower)"

49. Fill in the blank with this word: ""___ can you see ...?""

51. Pas ___ (dance solo)

53. Turban & this flower name share the same Turkish roots

55. ___ The Magazine (bimonthly with 35+ million readers)

59. Sports org.

60. Uzbekistan's ___ Sea

61. "Whipped Cream & Other Delights" frontman

64. Nesters

65. Wipe out

66. Young lady of Sp.

67. Time for playoffs

DOWN

1. So-called "white magic"

2. Uncle!'

3. French Bluebeard

4. What secondary recipients of e-mails get

5. Texter's 'Alternatively ...'

6. Subcompact

7. Spacewalks, to NASA

8. This unusual food falls from heaven for 40 years starting in Exodus 16

9. Part of an 800 collect call number

10. French possessive

11. Fill in the blank with this word: ""___ a Teen-age Werewolf""

12. He demonstrated that what Columbus had discovered was not 6-Down

13. Fill in the blank with this word: "___ shorthair (cat breed)"

14. Unscramble this word: oseserpn

21. Wow

23. African fox

25. Where Douglas MacArthur returned, famously

26. Mother of Xerxes I

29. Poultry place

30. Political extremists

31. Una ___ (old coin words)

35. Year in Septimius Severus's reign

37. They're spotted in Africa

38. University teacher

39. Medical inspiration?

40. Revenuers

46. Virgilian hero

47. Book size, in printing

50. Well-knit tales

52. Grow dark

54. Role for Ingrid

56. This won't hurt ___!'

57. Stadium sounds

58. Graceful bend

61. World War II Air Force commander ___ Arnold

62. Rock's Brian ___

63. Mao's mil. force

PUZZLE 79

64. Utter guilt, with "up"

65. Theatrical faint

DOWN

1. Thin, overseas

2. Win over

3. Particles in electrolysis

4. Levels

5. Words of encouragement

6. Elbowroom

7. U. of Maryland player

8. You might stick a knife in it

9. Kind of contract clause

10. Valentino title role, with "the"

11. *House that drains finances, slangily

12. Sigma follower

13. Wares: Abbr

19. Some sports cars

21. Springsteen's '___ Fire'

24. Repeated interjection in the Rolling Stones' "Miss You"

25. Vases

28. Violent behavior, to Brits

29. Guiding light

30. Utters

32. Nickelodeon's Kenan and ___

33. Poverty

34. Original Dungeons & Dragons co.

35. Some CBS forensic spinoffs

36. Prefix with -vert

37. Confronting boldly

41. Leading

42. Prefix with -stat

45. Prefix with sphere

46. Kansas' ___ River

47. Former TV host Sorkin

49. Renowned chair designer

50. Name placeholder in govt. records

52. The English translation for the french word: parsemer

54. Paris's ___ de Lyon

55. Waiting area announcements, briefly

56. Theme of this puzzle

57. S.A.S.E., e.g.

58. Wreath

ACROSS

1. Uninviting to a vegan

6. Al who sought the 2004 Democratic presidential nomination

10. Too pleased with oneself

14. Film score composer Morricone

15. Singer Green with multiple Grammys

16. 1956-57 Wimbledon champion Lew

17. Used to express farewell, this French word means "I commend you to God"

18. Stain looseners on washday

20. Nursery rhyme title fellow

22. ___ note

23. Fill in the blank with this word: ""Now I ___!""

26. Workup locales: Abbr.

27. Provoked

31. Fill in the blank with this word: "___ favor"

32. Passing reference in the "I Have a Dream" speech?

35. Pole tossed in competition

38. Volkswagen model

39. Tallinn natives

40. Detect Dan's cologne?

43. Hall-of-Fame basketball coach Hank

44. Played (around)

45. Winter weather, in Edinburgh

48. Nadia Boulanger's "La ___"

51. The English translation for the french word: monsieur

53. Preventive maintenance on a water barrier?

57. Competitor of "The 5th Wheel," in reality TV

59. Wipe, as a blackboard

60. Wild goose

61. Novus ___ seclorum (phrase on a dollar)

62. Oprah's 'Beloved' role

63. Weeds

PUZZLE 80

ACROSS

1. Slugfest

6. Stupid jerk

11. Two out of fifty?

14. Charge with another duty

15. Rattletraps

16. Fill in the blank with this word: "Bon ___"

17. Symbols on old manuscripts

18. Development developments

19. Where Mork and Mindy honeymooned

20. Make time for aerobics classes?

23. ___ go (the opposite of)

24. One-named Brazilian soccer star in the 2008 Time 100

25. The English translation for the french word: rial

27. Livestock feeds

29. CHICKEN LEGS

32. The English translation for the french word: Èmeu

33. 2nd-century date

34. Work started by London's Philological Soc.

35. Subject of this puzzle, born 10/11/18

39. Start of a laugh

40. Groups of animals, to biologists

41. Strauss's "___ Heldenleben"

42. Literary character whose name is said to mean "laughing water"

44. Without means of support?

48. Year of Philip I's birth

49. Windsurfers' mecca

51. Slugger Moises

52. Traitor of America

56. Suffix with ether

57. Fill in the blank with this word: "___ calculus (kidney stone)"

58. Word of parting

59. The "S" of R.S.V.P.

60. Witherspoon of "Vanity Fair"

61. Intervening, in law

62. The Gateway to the West: Abbr.

63. Fly catcher

64. Urge

DOWN

1. Seven-time N.F.L. East champions in the 1950's

2. Fresh start, as of a movie series

3. From behind: Lat.

4. Rout

5. Year the empress Octavia was murdered

6. They're unstressed

7. The English translation for the french word: houlette

8. Hobby with call signs

9. The English translation for the french word: humble

10. Fill in the blank with this word: "___ buco"

11. Virtual face

12. Abandons

13. NYSE buy

21. Kwame ___, advocate of pan-Africanism and the first P.M. of Ghana

22. Switzerland's Bay of ___

26. Woodstock supply

28. Wax theatrical

30. Sight from the Bering Sea

31. Weight-bearing bone

33. Neckcloth

35. Underage temptation, slangily

36. Two-time U.S. Open winner

37. Probed

38. Kansan or Coloradan

39. Thinking ...'

43. Writer Anais

44. Who often "did it" in a whodunit

45. Fill in the blank with this word: "___ Midgen, fellow student of Harry Potter"

46. Lawgivers

47. Washes with detergent

50. Make ___ for (support)

53. Transgresses

54. Yawning

55. Wheelchair access

56. Winding road shape

PUZZLE 81

ACROSS

1. Fill in the blank with this word: "___ Awards (annual prizes for African-American achievement)"

6. Western Indian

10. Twilights, poetically

14. This card deck's Minor Arcana has 14 cards in each suit; a page is between the 10 & jack

15. Unscramble this word: allh

16. Hollywood up-and-comer

17. Went furtively

18. Prefix with spherical

19. The Godfather' co-star

20. [Prediction:] Lance

Armstrong, at the end of the 2003 20-Across

23. Robertson of CNN

24. Where to see le soleil

25. Wanted-poster letters

28. When Parisians typically vacate

31. Struggle with

35. Writer Deighton

37. On ___ streak (winning)

39. Waterwheel

40. Start of the definition

43. Fill in the blank with this word: "___ Bowl"

44. Tutsi foe

45. Fill in the blank

with this word: "___ palm"

46. Calf muscle

48. Ukr. and Lat., once

50. Opium ___

51. One succumbing to 6-Down

53. Fill in the blank with this word: ""___ la la!""

55. Noted actor's underarms?

61. Puppeteer Tony

62. Subject, in Spain

63. Slippery ___

65. Fill in the blank with this word: "Family ___"

66. City in Judah

67. First saint canonized by a pope

68. French wave

69. Puccini soprano

70. Undergrad course, briefly

DOWN

1. Mail Boxes ___

2. Gross!

3. Uh-Oh! ___ (Nabisco product)

4. Fill in the blank with this word: ""___ intended""

5. Rare-earth oxide

6. It's you! What a surprise!'

7. Kind of bike

8. Early Mexican

9. Provoke

10. Prepares potatoes in a way

11. Jack of "Rio Lobo"

12. Fill in the blank with this word: "___ tide"

13. RR stop

21. Paint job finale

22. Poet whose works were set to music by Schumann, Strauss and Brahms

25. Take the part of

26. Fill in the blank with this word: "Artist Frida ___"

27. Fill in the blank with this word: ""He's ___ nowhere man" (Beatles lyric)"

29. Worrisome remark

by a surgeon

30. Soy foods

32. Threesome

33. Phoenix Suns head coach beginning 1996

34. Include as an extra

36. "Death in the Desert" writer, 1930

38. Times Square sign

41. Spherical

42. When you pull into Guadeloupe, you may have to change your dollars into this official currency

47. Wedding band, maybe

49. Gets the last of, as gravy

52. Trifles: Fr.

54. Forest ___

55. The English translation for the french word: grenier

56. Snow White's sister

57. Yao Ming teammate, to fans

58. Writer Janowitz

59. Whig's rival

60. Fill in the blank with this word: "___ and Span"

61. Pou ___ (vantage point)

64. Command level: Abbr.

PUZZLE 82

ACROSS

1. Suffers in the summer

7. Waterproof boots

13. Sandwich's title?

15. Multivitamin product

16. With 6-Down, common sight outside a school building

17. 'Talking Straight' author

18. Workplace watchdog, for short

19. Furniture protector

21. Zoologist's foot

22. Like some salts

24. Mil. pilot's award

27. Stone in Hollywood

28. Michael of 'Juno'

29. Uncle ___

32. Partied, so to speak

33. Fast gait

35. Pear variety

37. Northwest Terr. native

39. Roof worker

40. Scream for the Dream Team

42. Fill in the blank with this word: ""Physician, ___ thyself""

44. Fill in the blank with this word: "Conductor ___-Pekka Salonen"

45. Sportscaster Albert

46. 39-Down and others, for short

48. Alley org.

49. Place for some animal baiting

50. Fill in the blank with this word: ""Wham ___!""

53. It can cause a stir

54. Yarn

55. Out of action, in baseball lingo

58. Screen behind a church altar

61. One of 11 kings of Egypt

62. Teach, with "up"

63. Angers

64. Revolted

DOWN

1. The English translation for the french word: be-bop

2. Worker's demand

3. Violists' places: Abbr.

4. Three-time N.H.L. All-Star Kovalchuk

5. Parts of a house, in classifieds: Abbr.

6. Spain's Costa del ___

7. Not staying in one's lane

8. Went like a shooting star

9. House of ___

10. Mail Boxes ___

11. Mythical bird

12. Radical 1970s grp.

14. Umpire

15. Game, to Guglielmo

20. One with unusually fine hands?

22. E-6 officers in the U.S.A.F.

23. The Greek Letter Equivalent : R

24. Milan's Santa Maria ___ Grazie

25. Cleaving tools

26. 'You Can't Take It With You' director

28. Year in the reign of Antoninus Pius

29. The thing is what a lumberjack leaves behind; the place is where a politician speaks

30. 1942 Preakness winner

31. Stiller and ___ (comedy duo)

34. "It's about time!"

36. 8, on a phone

38. Zoo bosses

41. Speech enliveners

43. Fill in the blank with this word: "___ Onassis, Jackie Kennedy's #2"

47. Jumps (out)

49. Ending with wilde- or harte-

50. Go from ___ worse

51. Fugard's 'A Lesson From ___'

52. Intervening, in law

53. Syngman ___, first president of South Korea

54. Louisiana ___: Abbr.

55. Tulsa sch. named for an evangelist

56. Power ___

57. They often accompany logos: Abbr.

59. The English translation for the french word: Èmeu

60. Whack

PUZZLE 83

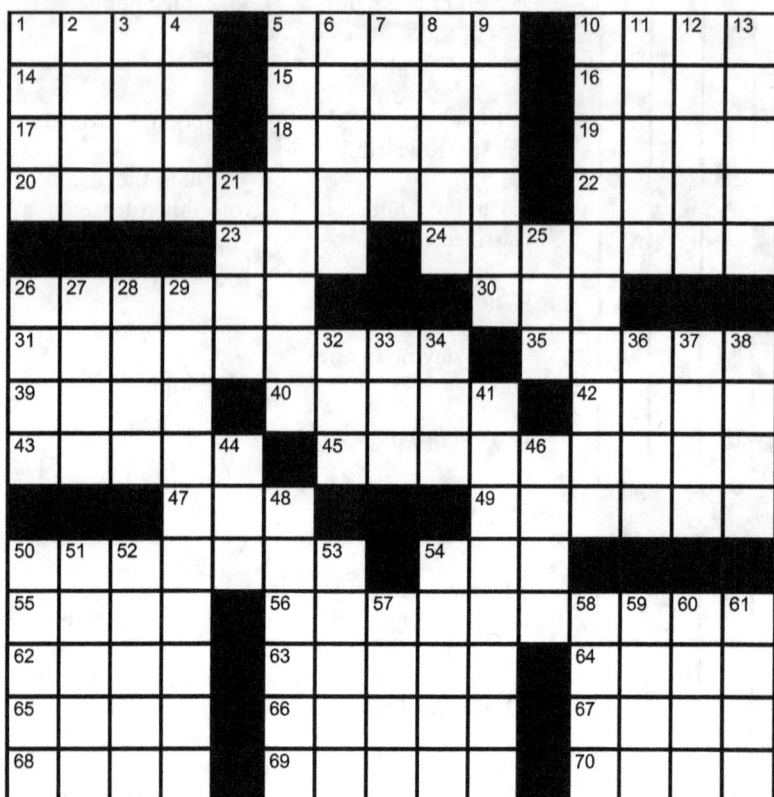

ACROSS

1. Workplace watchdog, for short

5. Fill in the blank with this word: "___ committee"

10. Vehicle pulled by a hoss

14. Unscramble this word: dere

15. Fill in the blank with this word: "___-Prayer"

16. Word of mock fanfare

17. Unusually excellent

18. Zero out

19. Some prosecutors: Abbr.

20. Laetrile source

22. The Soup ___

23. Sue Grafton's "___ for Lawless"

24. Stormy wintry blast

26. Words before "go"

30. Oh, sure'

31. Feign concurrence

35. Fill in the blank with this word: ""It Had to ___""

39. Massachusetts' ___ College

40. Fill in the blank with this word: "___ no"

42. Spanish ___

43. Food service giant based in Houston

45. Sealab inhabitants

47. Chem. formula for hydrogen isocyanide

49. Want to know

50. Sincere

54. Taro dish

55. Island in French Polynesia

56. Not taking no for an answer

62. Z ___ zebra

63. Miss Longstocking

64. Take ___ (swing hard)

65. Guy Lombardo's "___ Lonely Trail"

66. Prefix with -plasty or -gram

67. Very dark

68. Sharon of 'Dreamgirls'

69. Singer Rimes

70. Los ___ de los Muertos (Mexican holiday)

DOWN

1. Wroclaw's river, to Poles

2. Fat underwater creature

3. Mein ___

4. Whose woods these ___ think...': Frost

5. in an adroit manner

6. Waist removal regimens?

7. Rhyming with clasp, it's a clasp for a door or lid that's fastened with a padlock

8. Vegetable oil component

9. Stuffed mouse, maybe

10. Musial's nickname

11. Was sympathetic

12. In ___ (stunned)

13. The P.L.O.'s Arafat

21. See 26-Across

25. Music category

26. Mailing ctrs.

27. Fill in the blank with this word: "60's TV's ___ May Clampett"

28. Mai ___ (drinks)

29. Contemporaneous

32. War on Poverty agcy.

33. White House advisory grp.

34. Stuff

36. Common contraction

37. Bone: Prefix

38. The Beatles' "Back in the ___"

41. Common burger topper

44. Formula ___

46. Fill in the blank with this word: ""___ can't be!""

48. Kind of tunnel

50. The English translation for the french word: frîle

51. Spanish direction

52. Nancy's opposite number, once

53. Hanover's river

54. Fill in the blank with this word: "___ the Short, early king of the Franks"

57. Kraft Nabisco Championship org.

58. Wagered

59. About

60. Endangered Asian deer

61. 1974 Sutherland/ Gould spoof

PUZZLE 84

ACROSS

1. Plant with two seed leaves

6. Things to believe in

10. Warner Bros. cartoon name

14. Rodeo rope

15. French physicist noted for research on magnetism (born in 1904)

16. Obi accessory

17. Protection from enemy fire

19. Powerful kind of engine

20. The English translation for the french word: pleuvoir ‡ verse

21. Pro ___ (proportionately)

22. Fill in the blank with this word: "___ Tamid (synagogue lamp)"

23. Fill in the blank with this word: "___ Valley"

25. Troupe group

28. Academy Award song of 1947

31. Fill in the blank with this word: "Dress ___ (resemble)"

34. Fill in the blank with this word: "February ___ (Groundhog Day)"

35. Fill in the blank with this word: "___ Cologne (skunk of old cartoons)"

36. Comedians and parade directors?

40. Yak

41. Rapa ___ (Easter Island, to natives)

42. Sofer of soaps

43. Physical therapist's assignment?

47. Uncompromising leader

48. Leaf bisectors

53. What "y" might become

54. Nothing, in Nice

56. High-performance Camaro ___-Z

57. G.M., Ford and Chrysler

59. New Jersey town on Manasquan Inlet

61. Missing name in the tongue twister 'I saw ___ sawing wood ..'

62. Webb Pierce song "___ Know Why"

63. The Moselle flows into it

64. Hitting avg., e.g.

65. Unusual shoe spec

66. Secretary, at times

DOWN

1. Mild oaths

2. Middle of a famous palindrome

3. Joe ___

4. Mexican Indian

5. Toni Morrison's "___ Baby"

6. Sea breeze's heading

7. Banana ___

8. I never ___ man ...'

9. Trickiness

10. Some fairly difficult odds

11. Like virgin land

12. Rap's Dr. ___

13. Fill in the blank with this word: "___ Herbert, TV's Mr. Wizard"

18. Swift Malay boat

22. Operations ___ (Army position)

24. Fill in the blank with this word: "___ Nostra"

25. Supplementary: Abbr.

26. Nancy's opposite number, once

27. Fill in the blank with this word: "___-wolf"

29. Work ___

30. Treebeard, e.g.

31. Org. that used to bring people to court?

32. It has white plumage in winter

33. Site where trees are displayed

37. How many TV shows can be seen nowadays

38. Unscramble this word: ugy

39. Vexed

40. Fill in the blank with this word: ""A guy walks into a ___ "

44. Test result, at times: Abbr

45. Title girl in a 2001 French comedy

46. Fill in the blank with this word: "___ function"

49. Hindu sage

50. Train track beam

51. Fill in the blank with this word: ""___ chance!'"

52. Disperse

54. Way from Syracuse, N.Y., to Harrisburg, Pa.

55. Fill in the blank with this word: "___ dixit"

57. Roseanne's mom, on 'Roseanne'

58. The English translation for the french word: Isme

59. Joe-___ weed (herbal remedy)

60. Yip in a strip

PUZZLE 85

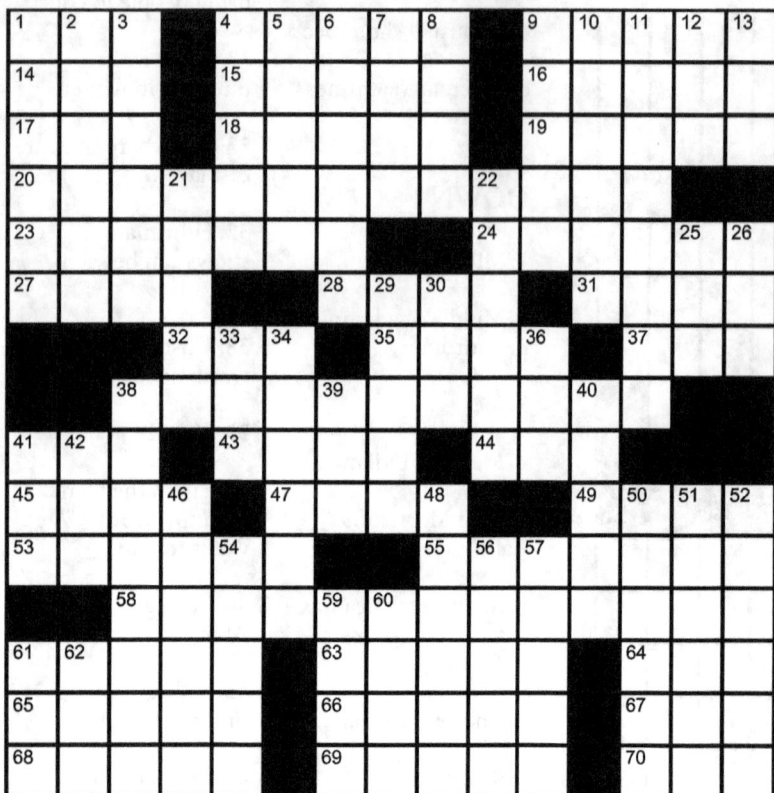

ACROSS

1. Unscramble this word: lto

4. The English translation for the french word: retomber

9. U.K. equivalent to an Oscar

14. Three ___ match

15. James ___ Garfield

16. Veldt sight

17. Educational cable network

18. They greet each other by pressing their noses together

19. Role-playing game, briefly

20. Heads-up in Ireland?

23. Toil

24. Actress de Ravin of "Roswell" and "Lost"

27. Present opener?

28. Fill in the blank with this word: "Director Vittorio De ___"

31. The English translation for the french word: silo

32. Part of C.P.I.

35. Workers' rights org.

37. Fill in the blank with this word: ""The Longest Day" director ___ Annakin"

38. 1965 hit by the performers suggested phonetically by the ends of 18-, 24-, 37- and 56-Across

41. Rembrandt van ___

43. Tony's portrayer on "NYPD Blue"

44. Soph. and jr.

45. Wife of Siva

47. This sac in a bird's egg is there to provide nourishment to the embryo

49. Band with the 1988 #1 hit "Need You Tonight"

53. Willingly taking

55. Stunned

58. Eliminates a blind spot, like a cosmonaut?

61. Faux pas

63. Madrid's ___ del Prado

64. Fill in the blank with this word: "Egypt's ___ Simbel"

65. verb remove the shucks from

66. Fill in the blank with this word: "___ Viejo (California city near Laguna Beach)"

67. Tiny ___

68. Fill in the blank with this word: ""If a body ___ body...""

69. Title role in a 1950s TV western

70. Shaker ___, O.

DOWN

1. Wish one could

2. Slot machine feature

3. Deli request

4. Female vampire

5. You can count on them

6. Walks about looking for prey

7. Hall & Oates: "___ Smile"

8. The ruler of Qatar is known by this 4-letter title

9. Shroud

10. Fill in the blank with this word: ""___ Unplugged" (hit 1999 album)"

11. Long and sharply pointed

12. Fill in the blank with this word: "February ___ (Groundhog Day)"

13. Word on a dipstick

21. Eye-related

22. Within earshot

25. QuÈbec's ___ d'OrlÈans

26. Years on end

29. Like many taste tests

30. Gulager of TV's "The Virginian"

33. Shad ___

34. Song from "The Music Man" with the lyric "What words could be saner or truer or plainer"

36. Fill in the blank with this word: ""A guy walks into a ___ "

38. Working together (with)

39. Ming of the Houston Rockets

40. "I Saw Her Standing There," vis-‡-vis "I Want to Hold Your Hand"

41. Fill in the blank with this word: "___ Pictures (old studio)"

42. Trap

46. Make a member

48. Like some cousins

50. The Jets retired his #12

51. Rapper with an MTV show

52. Shot putters' supplies?

54. See 70-Down

56. Weapons check, in brief

57. Not give ___

59. Wrangle

60. Anklebones

61. The English translation for the french word: DSM

62. Revolutionary Guevara

PUZZLE 86

(Crossword grid)

62. Unit of loudness

63. City in Judah

64. Fill in the blank with this word: "___ Day"

65. Former baseball commissioner Bowie ___

66. Saxophonist Zoot

27. What 17-, 23-, 37- and 51-Across may demonstrate?

29. Fill in the blank with this word: "___ law (old Germanic legal code)"

30. Sends back

32. Red as ___

33. Fill in the blank with this word: ""I swear I ___ art at all": "Hamlet""

DOWN

1. Iraq's ___ City

2. To me, to Mimi

3. Tracks

4. Bid

5. Part of Nasdaq: Abbr.

6. Play-___

7. Thought: Prefix

8. Fill in the blank with this word: "Alumni ___: Abbr."

9. Not put off

10. Wildcats' org.

11. Fill in the blank with this word: "___ prof."

14. Everything, to a lyricist

15. Manchurian border river

17. Fill in the blank with this word: ""Let's Get ___" (1973 #1 hit)"

21. Skeleton parts

23. Explosion sound

25. The English translation for the french word: Èquiper

26. Island near Quemoy

35. Corcoran of 'Bachelor Father'

38. Bonny young girl

42. Adam Dalgliesh's creator

45. Under the covers

47. Unpopular spots

49. Vols' school

50. You'll smile big if you know this 4-letter word for a long, thick piece of lumber used to support a roof

51. Soprano Berger

52. Expy.

53. Where Pearl City is

55. Sari-clad royal

56. In ___ (stuck)

57. Woolen caps

60. Fill in the blank with this word: "___-i-noor diamond"

ACROSS

1. Fill in the blank with this word: "___-masochism"

5. Fill in the blank with this word: ""___, I do believe I failed you" (opening of a 1998 hit)"

9. Rossini's '___ voce poco fa'

12. I ___ the opinion...'

13. Lays down the lawn

14. Lip-___ (fakes singing)

16. Small sharks

18. Fill in the blank with this word: "___ a hatter"

19. Meet, as expectations

20. Available from, as a product

22. Top of the ___ (Midtown observation deck)

24. Wide-headed fastener

25. First black N.F.L. Hall-of-Famer ___ Tunnell

28. Comparable to a rose?

31. Fill in the blank with this word: "___ minÈrale"

34. Muslim judge of North Africa

35. 39-Down and others, for short

36. TD scorers

37. Sport-___ (vehicle)

38. Some winter wear

39. Singer Des'___

40. Rose ___ rose...'

41. Beethoven's 'Sinfonia ___'

42. Striking end

43. Ye Olde Trip To Jerusalem claims to be the oldest of these in Britain, so bottoms up to it

44. Famed statement by 67-Down

45. Up

46. Spear

48. Topic: Abbr.

50. Italian for "beautiful singing"

54. Treat roughly

58. TV's Gray and Moran

59. Goya masterwork, with "The"

61. Starts a pot

PUZZLE 87

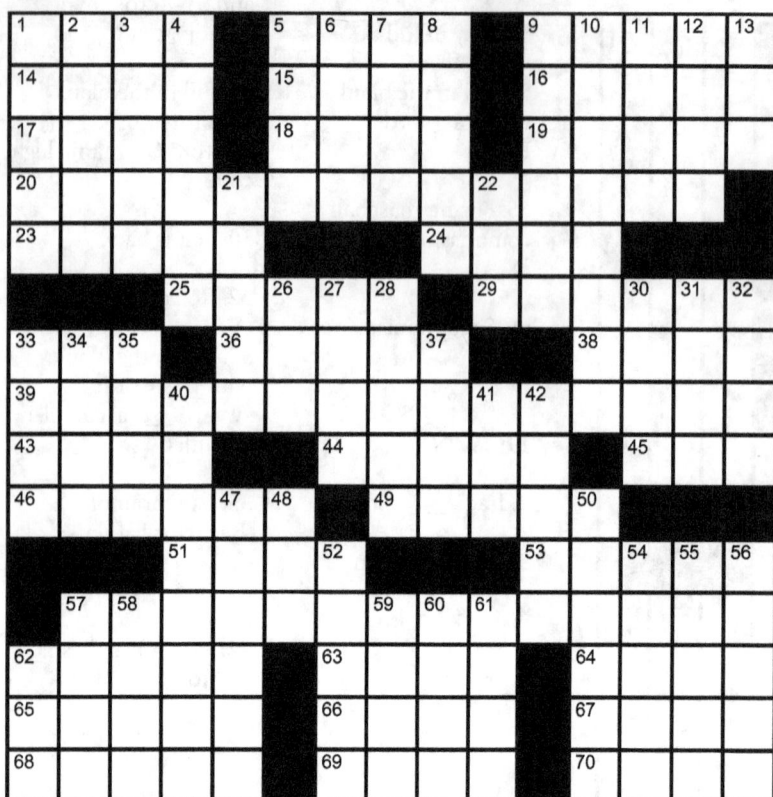

ACROSS

1. Fill in the blank with this word: "Emmy-winning actress ___ de Matteo"

5. Old N.Y.C. club said to be the birthplace of punk

9. L'___ du Tour (annual cycling event)

14. Finish this popular saying: "If a job is worth doing it is worth doing_____."

15. Vex

16. Ranger, e.g.

17. Simple rhyme scheme

18. Fill in the blank with this word: ""Who has seen the wind? Neither ___ you": Rossetti"

19. Five, on a gunslinger's gun

20. Printer's activity

23. N.J. post

24. Whoop

25. Hose woes

29. Oddball

33. What a guitar may be hooked up to

36. Declarer

38. North Carolina's Cape ___

39. Poem by 39-Across about breakfast in bed

43. Tribe in Manitoba

44. Wooden pail part

45. Secure online protocol

46. Below-the-belt

49. Troy, in poetry

51. Critic, at times

53. Unpleasant

57. George C. Scott movie with a rock band namesake, with "The"

62. The English translation for the french word: troupeau

63. Unscramble this word: liar

64. Sci-fi princess

65. On in years

66. Poet ___ Wheeler Wilcox

67. Not well

68. Fill in the blank with this word: "Australia's ___ Rock"

69. Fill in the blank with this word: "___ the Impaler"

70. 'Waiting for the Robert ___'

DOWN

1. Tower over

2. Disprove

3. Veldt sight

4. Fill in the blank with this word: ""___ Unplugged" (hit 1999 album)"

5. Crepe paper feature

6. Fill in the blank with this word: "___ cherry"

7. Unappetizing fare

8. Word that follows the start of 21-, 26-, 43- and 50-Across

9. Strand, in a way

10. Autobiography of 47A

11. Sam Shepard's "___ of the Mind"

12. I. M. and Mario

13. Verdi's "___ tu"

21. Slangy turndown

22. Repugnant exclamation

26. Prefix with fauna

27. Porter ___, former C.I.A. director

28. Three more than quadri-

30. Some wines

31. The English translation for the french word: estrade

32. Straight: Prefix

33. Yiddish writer Sholem

34. TV character first seen on "Happy Days"

35. Warm-up

37. The English translation for the french word: rial

40. Its valley is fabled in song

41. Year in Nero's reign

42. Israel's Bank ___

47. They're not allowed to travel

48. "Uh-huh"

50. Ear of Indian corn

52. Prelate's title: Abbr.

54. Tackle box item

55. Token taker

56. Fill in the blank with this word: "Belgian violin virtuoso Eugene ___"

57. Every 12 mos

58. Way it's done

59. Word that can follow the ends of 17-, 26-, 43- and 58-Across

60. City on the Gulf of Aqaba

61. Unscramble this word: ladg

62. Fill in the blank with this word: "___ one-eighty"

PUZZLE 88

ACROSS

1. Triumphant cries

5. W.W. II service acronym

9. "I Saw Her Standing There," vis-‡-vis "I Want to Hold Your Hand"

14. Fill in the blank with this word: "E. ___"

15. Anklebones

16. The English translation for the french word: tablier

17. Happy Days' cast member

19. Fox premiere of August 2003

20. Fill in the blank with this word: ""___ ma"

21. relating to or involving money

23. The Era of ___ (1964-74 Notre Dame football)

25. Wards (off)

26. How the perfect game is shown on the scoresheet

32. Fill in the blank with this word: "___ button (Facebook icon)"

33. Up-to-the-minute

34. The English translation for the french word: point par pouce

37. Struggles

38. Mark, as a survey box

39. Not his'n

40. Sue Grafton's '___ for Innocent'

41. Heretofore

44. Vast

45. 1965 film starring 55-/17-Across

47. Reference book abbr.

49. Inits. on old typewriters

50. Pass

54. Six-time Grammy winner Mary J. ___

58. ___ 300 (short-lived Apple laptop)

59. Impermanent

61. Vice ___

62. Victory: Ger.

63. Rock's Motley ___

64. Fill in the blank with this word: "Dan ___, former N.B.A. star and coach"

65. Molokai and Maui: Abbr.

66. Fill in the blank with this word: "Amerada ___ (petroleum giant)"

DOWN

1. Fill in the blank with this word: "Banda ___ (2004 tsunami site)"

2. Fill in the blank with this word: ""Que ___ es?" (Spanish 101 question)"

3. Fill in the blank with this word: ""___ volat propriis" (Oregon's motto)"

4. Extra costs of smoking and drinking

5. The English translation for the french word: OMC

6. ___ The Magazine (bimonthly with 35+ million readers)

7. Zoological wings

8. Pres., to the military

9. Be on deck

10. Landmark near the pyramids of Giza

11. Song from Sondheim's 'Passion'

12. Swingers in a saloon

13. Suffix with consist

18. Zeppo, for one

22. Polio vaccine pioneer

24. Singer ___ Rose

26. Year Plutarch was born

27. Unlucky number for Caesar?

28. Sporty Jaguars

29. Some horizontal lines

30. Singer Aguilera, self-referentially

31. The name of this gas comes from the Greek for "strange"

34. Sparrow player in films

35. Fill in the blank with this word: "___-dieu"

36. Go up: Abbr.

39. Eponymous doctor with a maneuver

41. Light

42. Not optional: Abbr.

43. Writing: Abbr.

45. Fill in the blank with this word: "___ fly"

46. Highly toxic pollutants

47. Renowned chair designer

48. Winter Palace residents

50. Fill in the blank with this word: "Dolly ___ of "Hello, Dolly!""

51. Who's there?' answer

52. Hindu titles

53. I Lost It at the Movies' writer Pauline

55. Fill in the blank with this word: ""Able was ___...""

56. Zoo attractions

57. Vous ___ ici'

60. Riot-stopping grps.

PUZZLE 89

ACROSS

1. Fill in the blank with this word: "___ spell"

5. Vessel with a load

9. Mandela's presidential successor

14. Mountain pool

15. When Othello decides he wants to poison Desdemona, this villain suggests that he strangle her instead

16. When you pull into Guadeloupe, you may have to change your dollars into this official currency

17. You'll use up 3 vowels playing this word that means toward the side of a ship that's sheltered from the wind

18. What an A is not

19. Routine

20. A state symbol of Maryland

23. Bond villain in "Moonraker"

24. Radical 1970s grp.

25. Wild-riding squire of "The Wind in the Willows"

28. The N.Y. Cosmos were in it

30. Globe: Abbr.

33. Fill in the blank with this word: ""I earn that ___": "As You Like It""

34. The Washington Post March' figure

35. Fill in the blank with this word: "___ Maria"

36. Story-filled magazine since 1922

40. Strauss's "___ Heldenleben"

41. Thundering

42. Fill in the blank with this word: "___ prius (trial court)"

43. The Beatles' ___ Pepper

44. The red-spotted type of this salamander is one of the most common in the U.S.

45. Symbol for a difficult ski run

47. Tokyo, once

48. Taco stand add-on, in brief

49. Purpose of Krzysztof's travels

56. Fill in the blank with this word: ""You ___ say that""

57. The blond Monkee

58. Fill in the blank with this word: "___ la Douce"

59. Sammy Kaye's "___ Tomorrow"

60. the periodic rise and fall of the sea level

61. Unscramble this word: vrya

62. Fill in the blank with this word: "___ means possible"

63. Fill in the blank with this word: "___-a-porter"

64. Fill in the blank with this word: "1990s N.F.L. running back Curtis ___"

DOWN

1. Wild guess

2. Fill in the blank with this word: ""___ show you!""

3. Something good

4. Speech enlivener

5. Singer O'Connor

6. The English translation for the french word: calice

7. Fill in the blank with this word: "___ arch"

8. With 33-Across, anagrams and puns (or parts hidden in 17-, 24-, 44- and 51-Across)

9. Hindmost brain parts

10. Anatomical sac

11. Italian emporium ending

12. City on the Rhein

13. Map abbr.

21. Malay Peninsula's Isthmus of ___

22. Exceptional rating

25. Tiny biters

26. Make ready to sail again

27. Twit

28. Yogurt type

29. Violinist Leopold

30. Zeno, notably

31. Belarus port

32. Member of a strict Jewish sect

34. Fill in the blank with this word: "Author C. P. ___"

37. In plain English

38. Two-tone treats

39. On the sidelines

45. KFC order

46. Fill in the blank with this word: ""___ note to follow ...""

47. Mystery writer Stanley

48. Fill in the blank with this word: "___-robe (Calais closet)"

49. The English translation for the french word: poney

50. Make ___ check

51. WWW address starter

52. Fill in the blank with this word: "___ dire"

53. Fill in the blank with this word: ""___ (So Far Away)" (1982 hit by A Flock of Seagulls)"

54. TV actor Katz

55. Votes against

56. Young carnivore

PUZZLE 90

ACROSS

1. Delaware Indian whose name is French for "a friend"

6. Triangular sails

10. Fill in the blank with this word: "Chocolat au ___"

14. Make ready to sail again

15. Fill in the blank with this word: "___-Day"

16. You never had ___ good!'

17. Tiny bit

18. Certain supermarkets, for short

19. Israeli resort city

20. How long 25-Across was 41-Down before being noticed and fixed

23. Go ___ some length

24. "Venice Preserved" dramatist Thomas

25. Used in kendo, the shinai is a sword made of this plant

28. Literature Nobelist ___ Fo

31. Family name suffix in taxonomy

32. Polar irregularity

33. Renault, e.g.: Abbr.

36. Rafael Nadal specialty

40. Fill in the blank with this word: "___ Kosh B'Gosh"

41. Fill in the blank with this word: ""What ___ boy am I!""

42. Part of 16-Across: Abbr.

43. Mythical eponym of element #41

44. They were named after Henry's son

46. Flavor

49. Standoffish

50. Apology #1

56. Largest employer in Newton, Iowa, until 2006

57. Some reddish deer

58. Yucat

60. Chase of "Now, Voyager"

61. Fill in the blank with this word: "___ Aarnio, innovative furniture designer"

62. Weapons check, in brief

63. Fixed at an acute angle

64. Sound of a leak

65. Bridge opponents

DOWN

1. Tail: Prefix

2. Nine, in Nice

3. The Ponte Vecchio crosses it

4. Disease-causing bacteria

5. Fill in the blank with this word: ""___ kick from champagne...""

6. Floorboard supporter

7. William who wrote "The Dark at the Top of the Stairs"

8. Wally's TV brother, with "the"

9. Writer's Market abbr.

10. Wasn't straight with

11. Fill in the blank with this word: "Counselor-___"

12. Warwick's "___ Little Prayer"

13. Hotsy-___

21. Ming of the Houston Rockets

22. Hold the rocks'

25. Fill in the blank with this word: "___-Honey (candy name)"

26. Uproars

27. "Happy Days" fellow

28. Robinson Crusoe' author

29. Youngest player to join the 500-homer club

30. Soldier

32. Uppity type

33. Zoo feature

34. With: Abbr.

35. Fill in the blank with this word: "___ Grand (supermarket brand)"

37. Twosomes

38. Fill in the blank with this word: "1987 #1 hit "Here ___ Again""

39. Believes

43. Sampler sample

44. Wright wing?

45. Fill in the blank with this word: ""Who ___?""

46. Weigh station sights

47. Skating jumps

48. Rich dessert

49. Lhasa ___ (Tibetan dogs)

51. Fill in the blank with this word: ""So ___ to you, Fuzzy-Wuzzy": Kipling"

52. Fill in the blank with this word: "___ Southwest Grill (restaurant chain)"

53. Suffixes with ballad and command

54. Go on a vacation tour

55. user-friendly (similar term)

59. U.S.N. officers

PUZZLE 91

ACROSS

1. Woe ___ them that call evil good': Isaiah

5. Like many gardens

10. Yemen's capital

14. Flies away

15. Toughen, as to hardship

16. Speaker of note

17. Weirded-out feeling

20. Coach's strategy

21. ___ Sea (Italy/Greece separator)

22. The English translation for the french word: cÈ

23. The Pointer Sisters' "___ Excited"

24. Fill in the blank with this word: "___ Ababa"

27. March Madness feature

31. Venomous snake

32. This meat is also called kid & when curried is a Jamaican specialty

33. The English translation for the french word: pod

34. Attempts to climb a mountain range?

38. Soprano Christiane ___-Pierre

39. Tend to a hole

40. In an infamous "Cheers" episode, this character's husband, Eddie LeBec, was run over by a Zamboni

41. Resort island northeast of Sydney

44. Sales slips: Abbr.

45. Suffixes with ballad and command

46. Fill in the blank with this word: "Electric ___"

47. Reaction from one who has a bone to pick?

50. Juvenal, for one

55. Bit of attire for a business interview, maybe

57. Mulling

58. Irish P.M. ___ de Valera

59. Typewriter type

60. Useful Latin abbr.

61. "Nadja" actress L

62. Go ___ some length

DOWN

1. Naut. law enforcers

2. Wildcats' org.

3. Loc. of some devils

4. Douay prophet

5. You'll get a kick out of them

6. Temple architectural features

7. Former baseball commissioner Bowie ___

8. We'll teach you to drink deep ___ you depart': Hamlet

9. Ravage

10. Trial position, for short

11. Whose woods these ___ think...': Frost

12. Thatching palm

13. Fill in the blank with this word: "Alumni ___: Abbr."

18. Some footnotes, for short

19. Sub ___ (secretly)

23. Raider Carl

24. Southwest land

25. What is the capital of this country - Senegal

26. Finish this popular saying: "Eat, drink and be merry, for tomorrow we_____."

27. Fill in the blank with this word: "___ Dame"

28. Unscramble this word: atarp

29. Slaves

30. Writer ___ St. Vincent Millay

31. TV music vendor

32. Vexes, with "at"

35. Incapable of being detected, in a way

36. Year St. Augustine was born

37. World War II Air Force commander ___ Arnold

42. Clockwise

43. Now see ___!'

44. The English translation for the french word: rÈtine

46. John H. ___, Jackson's first Secretary of War

47. Winnie-the-Pooh's Hundred ___ Wood

48. Wipeout

49. Low-cost home loan corp.

50. Reindeer-herding people

51. Fill in the blank with this word: ""___ Man," Emilio Estevez movie"

52. Pack ___ (quit)

53. Fill in the blank with this word: "Director Vittorio De ___"

54. The English translation for the french word: tÈtine

56. Fill in the blank with this word: "___-d'Oise (French department)"

PUZZLE 92

ACROSS

1. The King ___'

5. Datebook abbr.

9. Greasy ___

14. Some Bourbons, par exemple

15. Fill in the blank with this word: "___ John"

16. Le ___, France

17. Start of a Yogi Berra quote

19. 1972 #2 hit for Bill Withers

20. Like a sunken treasure?

22. Traditional Sunday fare

23. Deferential

24. Response to a compliment

25. Turing test participant

26. Xerophyte's home

28. Year Vespucci sailed to the New World

30. Red ___

31. Unpaid interest?

35. Video store category

38. Pou ___ (vantage point)

39. Older couple's home, often

40. Old radio's Fibber ___

41. Bright, to Brecht

42. Comparable in years

43. Overcast

45. Genealogist's abbr.

47. Fill in the blank with this word: "___-da (pretentious)"

50. File on an iPod

51. Malory's 'Le ___ d'Arthur'

53. See 7-Down

57. Very cold

58. Vector around the equator

59. Whittle away

60. Rat-___

61. Fill in the blank with this word: "___ gum"

62. Vespers preceder

63. W.W. II vessels

64. Veteran journalist ___ Abel

DOWN

1. Fill in the blank with this word: "Bust ___ (laugh hard)"

2. Like some cigarettes

3. Stripping

4. "Amen!": 3 wds.

5. Like some professors

6. In the ___ Colony' (Kafka story)

7. Workers in chalk

8. Tale of a cease-fire?

9. Oscar-nominated actress for "Leaving Las Vegas"

10. Grapevine exhortation

11. British runner Steve

12. an abalone found near the Channel Islands

13. Wanting

18. White House's ___ Room

21. Fill in the blank with this word: "'___ Haw'"

26. 1867 book subtitled "Kritik der politischen

27. Line score letters

28. Vegas's ___ Grand

29. Fill in the blank with this word: "___ 20, Natl. Games Day"

32. Stunning

33. Some undergrad degs

34. Wall Street earnings abbr.

36. Multiply like an amoeba

37. Most miserable hour that ___ time saw': Lady Capulet

44. Fill in the blank with this word: "Capitol-___ (music company)"

45. The Everly Brothers' "All I Have to ___ Dream"

46. French department in the Pyrenees

47. Wearer of three stars: Abbr.

48. Thackeray's "Vanity Fair: A Novel Without ___"

49. French actor Alain

51. Pitchers' gloves

52. Zhou ___

54. St. ___ (malt liquor brand named after an Irish nun)

55. Fill in the blank with this word: ""I earn that ___": "As You Like It'"

56. Fill in the blank with this word: "Bronti's "Jane ___""

PUZZLE 93

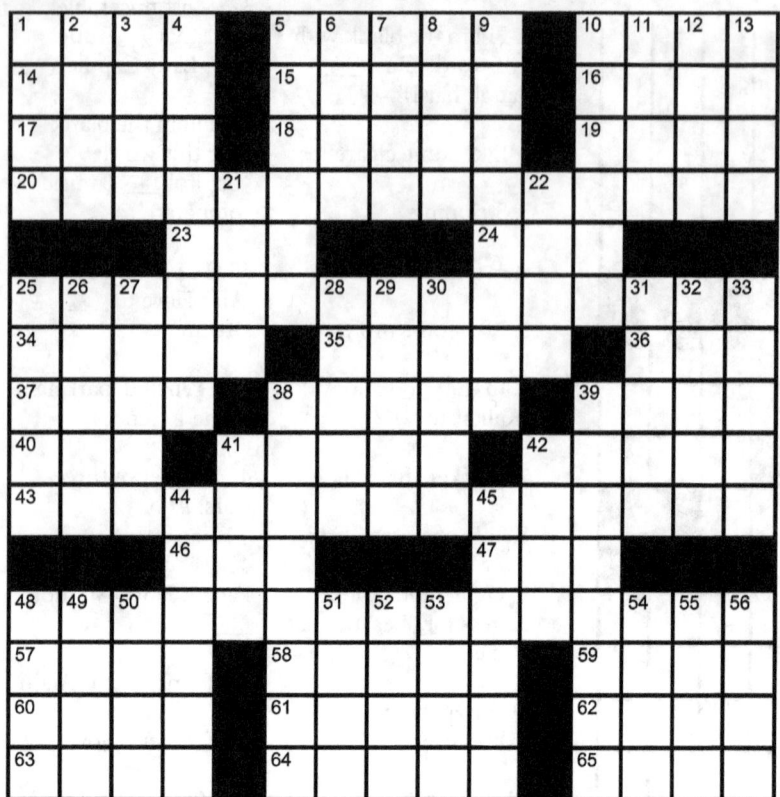

ACROSS

1. Wroclaw's river, to Poles

5. Club

10. Relative of "Oh, no!"

14. Old English bard

15. Wordsworth's Muse

16. Mythical mother of the Titans

17. ___ wrestler

18. Fond ___, Wis.

19. The English translation for the french word: germe

20. Novel featuring Adela Quested

23. French shooting match

24. The Equality State: Abbr.

25. One working for a flat fee?

34. Wire: Abbr.

35. Old card game

36. W. Sahara neighbor

37. Transfer and messenger materials

38. Rose

39. Harry Hershfield comic "___ the Agent"

40. Sue Grafton's '___ for Evidence'

41. Wake Island, for one

42. Repetition for rhetorical effect

43. Why the dog caught a cold?

46. Fill in the blank with this word: ""___ la la!""

47. Silver ___

48. Ida Lupino, e.g.

57. Way from Syracuse, N.Y., to Harrisburg, Pa.

58. Words to remember

59. Fill in the blank with this word: ""Zip-___-Doo-Dah""

60. About

61. Slippery as ___

62. Word to waiters

63. Cannon shot on a set?

64. What a tragedy!'

65. Fill in the blank with this word: "___ sample"

DOWN

1. Greece's Mount ___

2. Large bra feature

3. Tomato type

4. Followers

5. Los Angeles's ___-Sinai Medical Center

6. Fill in the blank with this word: "Designer ___"

7. The English translation for the french word: tempĺte

8. Tiers ___ (French commons)

9. Cause of delirium in farm animals

10. The English translation for the french word: lait de poule

11. Vexed

12. Gas: Prefix

13. Lady of Spain

21. Victory: Ger.

22. Month after Nisan

25. Prelate's title: Abbr.

26. Fill in the blank with this word: "___, meenie, miney, mo"

27. Fill in the blank with this word: "As ___ resort"

28. From bottom ___

29. Whack-___

30. Letter-shaped fastener

31. Form into an arch, old-style

32. Hold the rocks'

33. Questionable cradle location

38. One of the Virgin Islands

39. Causing change

41. Fill in the blank with this word: ""___ Flux" (Charlize Theron film)"

42. Vivacious

44. Wok concoction

45. Scotland's Firth of ___

48. Fill in the blank with this word: "___ Gailey of "Miracle on 34th Street""

49. Response to a compliment

50. Watercolorist ___ Liu

51. Prefix with sphere

52. Ways: Abbr.

53. Fill in the blank with this word: "___ lot (gorged oneself)"

54. Schwalm-___ (German district)

55. The English translation for the french word: sexy

56. Fill in the blank with this word: "___ time limit"

PUZZLE 94

ACROSS

1. What were the names of the 3 Cartwright sons?

5. Flightless bird: Var.

9. Kind of bike

13. Liniment ingredient

16. Seraph of S

17. Rage

19. Mare : horse :: ___ : sheep

20. Zoot-suited, say

21. Learner's ___

22. Second to ___

24. The World Book illustration for this occupation has a man with a cutlass, daggers, a pistol & a boarding ax

27. Fill in the blank with this word: "" ___ was saying Ö""

28. Rose-red dye

30. O-___-O (brand of sponge)

31. Quintillionth: Prefix

32. Fill in the blank with this word: ""You've ___ Mail""

34. Metalsmith's tool

37. Numbskull

39. Stars of "The Breakfast Club" and "St. Elmo's Fire," collectively

41. Clearasil target

42. Seal's opening?

43. Pride : lion :: clowder : ___

45. Cello feature

49. The English translation for the french word: sautiller

50. New York's ___ Bay Park

53. Role in "Son of Frankenstein"

54. Some college donors

56. Wedge-shaped inlet

58. Fill in the blank with this word: "Cambodia's ___ Nol"

59. Lawyers' requests at trials

63. Fill in the blank with this word: "Aid and ___"

64. Showers of purchases

65. Your throat might be this from yelling

66. The Beatles' "Back in the ___"

67. What ___' (1996 Sublime hit)

DOWN

1. Sunflower seed, botanically

2. Korean carmaker

3. Somme place

4. It's higher on the hwy.

5. Egyptian for "be at peace"

6. Fill in the blank with this word: "___ of Capricorn"

7. Fill in the blank with this word: "___ a kind (pair)"

8. U.K. counterespionage agcy.

9. Asian goat

10. The English translation for the french word: vivant

11. Chanel fragrance for men

12. Unscramble this word: toemnni

14. RMN opponent

15. Computer type

18. GM: "___ the USA in your Chevrolet"

23. Pizza slice, usually

25. One of Judaism's four matriarchs

26. Wing-shaped

29. Informal denial

31. What a shrug may indicate

33. Bygone carrier

35. Young fellow

36. Mariner ___ Ericson

37. Hi-tech auto device

38. Three-quarters of M

39. Popular vacation locale

40. Last film directed by Howard Hawks

44. Wheel of Fortune et al.

46. Viewing with elevator eyes

47. Rely on

48. Fill in the blank with this word: "Cyclotron inventor ___ Lawrence"

50. Kung ___ chicken

51. Twilights, poetically

52. Unlikely grant giver

55. Tiny bit

57. You can bank on it

60. Fill in the blank with this word: "Disco ___ (character on "The Simpsons")"

61. Fill in the blank with this word: "___ in Thomas"

62. Retailer with stylized mountaintops in its logo

PUZZLE 95

ACROSS

1. Year that Chaucer died

4. Yellow finch

9. Cable TV giant

12. Trust fund babies, often

14. Fill in the blank with this word: ""I swear I ___ art at all": "Hamlet""

15. Play ___ with (damage)

16. The start of 17-Across or 11- or 33-Down

18. WSW's reverse

19. People taking les examens

20. Laying hold of

22. One in debt?

24. Zodiacal sign for late March & early April

25. Writer Josephine

26. The English translation for the french word: pleuvoir ‡ verse

28. Where to split hairs?

31. Never idle

33. Tiber tributary whose name means "black"

34. Fill in the blank with this word: ""American Idol" winner ___ Studdard"

37. In ___ of

38. Saudi Arabian currency

39. Fill in the blank with this word: "Dress ___ (resemble)"

40. Small river craft: Var.

42. Vivacious person

44. Thread: Prefix

45. QVC alternative

48. High-altitude home

50. Supports a foundation

52. Transient things

55. Fill in the blank with this word: "A ___ (kind of reasoning)"

56. Mai ___

57. Kindly

59. Sta. purchase

60. Massenet opera set in Castile

61. Fill in the blank with this word: ""What ___ to do?""

62. Smart ___ (wise guy)

63. One of a Western political family

64. Width measure

DOWN

1. Ear of Indian corn

2. She won an Emmy playing Miss Jean

3. The English translation for the french word: troupeau

4. Fill in the blank with this word: ""___, bro?""

5. Intuitive feelings

6. Spanish queen

7. Viscera

8. Bean

9. With 9-Down, group with a 1962 hit version of 39-/41-/ 42-Across

10. Hockey's ___ Smythe Trophy

11. Fill in the blank with this word: ""___ your pardon?""

12. Thermonuclear experiment of the '50s

13. Sir Frank ___, historian of Anglo-Saxon England

17. Mixer maker

21. Possible item in a window box

23. Hezbollah stronghold ___ Valley

27. Workweek letters

29. The Era of ___ (1964-74 Notre Dame football)

30. Francis Poulenc's "Le ___ masqu

31. Certain terrier, informally

32. Stinky

34. Was revolting

35. Zogby poll partner

36. Singles, e.g.

38. With 51-Across, wet-day wish

40. Like some profanity

41. Waive one's rites?

43. Gold-imitating alloy

45. All over

46. Unruffled

47. [See title, and proceed]

49. Tropical palm

51. Engine attachment

52. Jazz singer ___ James

53. Bundled units, in some product names

54. Shade of blue

58. On a blood test, a level of 60 or above of this, the "good cholesterol", is healthy

PUZZLE 96

ACROSS

1. View in northern Italy

5. First Chinese dynasty

9. Toni Morrison novel

13. Spell

14. Vital carrier

16. Support

17. Woodworking tool

18. Surly TV bartender

20. Engrossed

22. Cosmetics dye: Var.

23. "Venice Preserved" dramatist Thomas

24. Warship

25. "Fiddler on the Roof" setting

27. 1985: Jeff Daniels steps out of "The Purple Rose of ___"

28. Fill in the blank with this word: ""Am ___ risk?""

29. Like the rarest rhino

31. Administer

35. Unscramble this word: soht

37. Rock's ___ and the Dominos

39. Verbal assault

40. Fill in the blank with this word: ""You're ___ trouble!""

42. Surface anew

44. Muhammad ___

45. Kitty-cat

47. Like some cousins

49. Drosselmeier's title in "The Nutcracker"

52. Fill in the blank with this word: ""A merry heart ___ good like a medicine": Proverbs"

53. Comic Boosler

54. Gadget-laden

57. Venetian painter

59. The English translation for the french word: tempîte

60. go (the opposite of)

61. Mandela's presidential successor

62. They replaced C rations

63. Bone: Prefix

64. This Indian flat bread with a palindromic name is traditionally baked in a tandoor oven

65. Vitamin C

DOWN

providers, maybe

1. Dugout shelter

2. Wall-plastering material

3. *"Cheers!" ... or a hint to answering this puzzle's five starred clues

4. Tropicana Field locale, informally

5. Deli sandwich material

6. Message for a pen pal?

7. Vexed

8. Confounded

9. Congressional periods

10. Wrinkly fruit

11. Tropical vine

12. Quiz show host, often

15. an ethnic group living in Azerbaijan

19. Fill in the blank with this word: "Cahn-Styne's "___ My Girl""

21. Medieval romance tale

24. Candy ___ (Christmas decorations)

25. Spanish ayes

26. Chemistry Nobelist Otto

27. Singer with the 1993 multi-platinum album "Music Box"

30. Vice ___

32. a card with words or numbers of pictures that are flashed to a class by the teacher

33. Colombian city

34. Key of Brahms's Symphony No. 4: Abbr.

36. What a child may stand on

38. City near Grissom A.F.B.

41. Rich fertilizer

43. Per ___ (daily)

46. Fill in the blank with this word: "___ whale"

48. The scarlet letter, e.g.

49. I can't ___ satisfaction'

50. Stews

51. Relatives of "Gee whiz" and "Shucks!"

52. Tough-talking coach

54. Lack of vigor

55. Swiss artist Paul ___

56. Boot camp affirmative

58. Surveyor's dir.

PUZZLE 97

DOWN

1. Some riffraff

2. Playwright Jean

3. Wordless song: Abbr.

4. Decline an invitation

5. William Blake: "When the stars threw down their spears,/ And watered heaven with their _____"

6. Kill ___ killed' (desperate mantra)

7. The "S" of R.S.V.P.

8. Fill in the blank with this word: ""___ was in our lips and eyes": "Antony and Cleopatra""

9. Sartre's first novel

10. St. Augustine's language

11. Vivacious wit

12. Realtor's specialty, for short

14. Fill in the blank with this word: "___ Langer, who wrote "Philosophy in a New Key""

18. Suffix with material

22. Types with fat recording contracts

24. One of the Simpsons

26. 1988 Peter Allen musical

27. Use a surgical beam

28. Fill in the blank with this word: ""Look at me, ___...""

29. Humbug

30. Sorry soul

31. Metal joiner

32. Wilkes-___, Pa.

35. Unclutter

36. Elicits an "ick!"

38. Russian ruler: Var.

39. Where Hercules strangled a lion

41. Turkish inn

42. Happen to

44. Hollywood's Alan and Diane

45. Sam of 'Jurassic Park'

46. Classic Marx Brothers flick

47. Old laborers

48. Patience, e.g.

49. Fill in the blank with this word: ""___ be in England...""

50. Fill in the blank with this word: ""Young ___ Boone" (short-lived 1970s TV series)"

53. Writer Santha Rama ___

ACROSS

1. Fill in the blank with this word: "___ Friday's"

4. Milk: Prefix

9. What an A is not

13. Nesters

15. Birthstone for most Leos

16. Fill in the blank with this word: "Europe's Gorge of the ___"

17. No good

19. Gas, e.g.: Abbr.

20. Becomes peeved

21. Meet, as expectations

23. Low-altitude cloud

24. Pioneer geneticist

25. Tiny criticism

26. Boxer who wrote "Reach!"

29. Swiss money

32. Some Muppet dolls

33. Topper

34. The English translation for the french word: leurre

35. Shrewlike

36. Laughter, in La Mancha

37. Truman's nuclear agcy.

38. Cover ... or cover ___

39. Bogot· babies

40. Don Herbert's moniker on 1950s-'60s TV

42. Roseanne's mom, on 'Roseanne'

43. ___ 300 (short-lived Apple laptop)

44. Paris paper

48. 1958 #1 song with the lyric "Let's fly way up to the clouds"

50. Like men at stag parties

51. Scarlett O'Hara's home

52. Certain atomic X-ray emission

54. Way from Syracuse, N.Y., to Harrisburg, Pa.

55. Yo-yoing

56. Turn on an axis

57. Wrongful act

58. Vitality

59. Whirled records?

PUZZLE 98

ACROSS

1. Fill in the blank with this word: "___ Nostra"

5. Scale with five sharps: Abbr.

9. Stunning blow

14. Stuck in ___

15. Yesterday, in Italy

16. Luck Be ___ Tonight'

17. Unbuild

18. City in Judah

19. Passover's month

20. Ulysses S. Grant's anagrammatic advice regarding hangovers

23. Liberace wore them

24. Tone deafness

27. Shar-___ (wrinkly dog)

28. Fill in the blank with this word: "Dick Francis book "Dead ___""

30. Philosopher Mo-___

31. Police dept. employee

34. Young Republican of a 1980s sitcom

37. Sayers's Bunter, e.g.

39. Parisian article

40. Work in a refinery

41. In an unnatural way

44. Go-aheads

45. Wedding page word

46. Route for Ben-Hur

47. Where Switz. is

49. View

51. Victims of ostracism

55. Little laughter while still on the runway?

58. Shell competitor

60. Part of Popeye's credo

61. Fill in the blank with this word: ""I earn that ___": "As You Like It""

62. Wordsworth's Muse

63. One-named Brazilian soccer star in the 2008 Time 100

64. Wrist-elbow connector

65. This word referring to a Roman soothsayer has become a verb meaning "to predict"

66. Esquire in "Henry VI, Part 2"

67. Whipped up

DOWN

1. Vehicles on the links

2. Fill in the blank with this word: ""Laborare est ___" ("to work is to pray")"

3. Popular snack cake

4. Took something in at night?

5. Two-year stretches

6. Neoclassicist who painted the fresco "Parnassus"

7. Uzbekistan's ___ Sea

8. Iwo ___

9. Modern places for groups of groupies

10. Venusian, e.g.

11. Blockbuster rentals

12. Soprano Christiane ___-Pierre

13. Rembrandt van ___

21. Turnabout, slangily

22. Chimney-top nester

25. Nitrogen compound

26. They're high in Manhattan

28. Middle: Prefix

29. N.B.A.'s Nick Van ___

31. Fill in the blank with this word: "___ Zelnicek (celebrity's maiden name)"

32. Local theaters, in slang

33. Camping gear

35. Fill in the blank with this word: "Debussy's "Clair de ___""

36. Element used in radiation research

38. noun a flag having three colored stripes (especially the French flag)

42. Battle of ___ (1943 U.S./Japanese conflict)

43. Cat's assent

48. Winner of the first World Cup: Abbr.

50. Verdi baritone aria

51. Rich dessert

52. Kipling's wolf pack leader

53. Midwest and Plains states, e.g.

54. Stiff hairs

56. Fill in the blank with this word: "___ Taylor, co-host of "Make Me a Supermodel""

57. Twosome

58. Give __ break!'

59. Tulsa sch. named for an evangelist

PUZZLE 99

ACROSS

1. Rubaiyat' rhyme scheme

5. Break activity, perhaps

10. The last Mrs. Chaplin

14. Scientology's ___ Hubbard

15. Fill in the blank with this word: "___ Good Feelings"

16. Give ___ to (approve)

17. Traditional end of summer

20. What 1938's "The War of the Worlds" broadcast set off

21. Lane with lines

22. Otologist

23. Time ___ half

24. James of "Gunsmoke"

27. Prefix with fauna

28. Yago Sant'___ (wine brand)

32. With it

33. Fill in the blank with this word: "___ note (dictionary bit)"

35. Fill in the blank with this word: "___-80 (old computer)"

36. Ad line #3

39. One of the Cyclades

40. Take countermeasures

41. Fill in the blank with this word: ""___ could have told you that!""

42. Theodore Roosevelt Natl. Pk. setting

44. Pale ___

45. Square, in 1950s slang, indicated visually by a two-hand gesture

46. Supplementary: Abbr.

48. Heraldic fur

49. Old import tax

52. Do spadework?

56. Ottawa official

58. Sketch

59. Wagner's Tannh

60. Cell stuff that fabricates protein, for short

61. Times past

62. What your nose knows

63. Go-aheads

DOWN

1. Full of compassion

2. Without ___ of hope

3. Some fishing gear

4. Not yours

5. Fill in the blank with this word: "___ Errol, main character in "Little Lord Fauntleroy""

6. It's ___!' (speakeasy cry)

7. Water brand

8. Scoot

9. Divisions of a mark

10. Poster stock

11. Small and insignificant

12. Soul singer Hendryx

13. Wing: Abbr.

18. Person with lots to sell

19. Year that Augustus exiled Ovid

23. Fill in the blank with this word: "___-garde"

24. Garson ___, writer and director of Broadway's "Born Yesterday"

25. This 5-letter word can refer to an event in Genesis or to what you do in supplying too much fuel to your carburetor

26. Fill in the blank with this word: "___ good example (shows the proper way)"

27. Fill in the blank with this word: ""Cold ___" (Foreigner hit)"

29. Prelate's title: Abbr.

30. Fill in the blank with this word: ""Goodnight, ___" (#1 hit of 1950)"

31. Words after ugly or guilty

33. One of a Western political family

34. Summer wind in the Mediterranean

37. Provide a segue for

38. Rescuee's cry

43. Black and Valentine

45. Some brews

47. Palme ___ (Cannes award)

48. Forcefulness

49. To do this is to stare at someone desirously

50. Rope fiber

51. Rating of a program blocked by a V-chip

52. Villainous resident of Crab Key island

53. ___ Island (location near Portland, Maine)

54. The Beatles' "Penny ___"

55. They, in S

57. Work started by London's Philological Soc.

PUZZLE 100

ACROSS

1. Certain constrictor

6. 'Today ___' (California morning show)

10. Xerox setting: Abbr

13. What Spanish athletes go for at the Olympics

14. Excerpts

15. Width measure

16. 17- and 60-Across and 11- and 35-Down

19. Get out of shape?

20. Fill in the blank with this word: "___ occasion (never)"

21. Showy flower of the iris family

22. Moderated

24. Fancy salad ingredients

29. The English translation for the french word: roseau

30. Hydrocarbon suffixes

31. Seasoned singer?

39. Fill in the blank with this word: "___ use (worthless)"

40. Nagy of Hungary

41. Fission boat?

48. Fill in the blank with this word: "___ rasa (clean slate)"

49. 'Vette option

50. Spy Aldrich ___

51. a rack with hooks for temporarily holding coats and hats

55. Madison Avenue's "loneliest guy in town"

60. Pale ___

61. Le ___, France

62. Psychiatrist/author R. D. ___

63. Fill in the blank with this word: "___ Tamid (synagogue lamp)"

64. Where Loews is "L"

65. The plural of the word spy

DOWN

1. Truman's nuclear agcy.

2. With a "K" at the end, it's a wooden toy; without, a group of legislators who voted together

3. With 121-Across, part of an afternoon repast

4. Ancient Greek sculptor of athletes

5. Of a large artery

6. 1972 Staple Singers #1 hit

7. Vardalos of the screen

8. Orderly supervisor, maybe: Abbr.

9. Fill in the blank with this word: "___ was saying Ö"

10. Two-point throw in horseshoes

11. PC-linking program

12. Fixes, as some fairways

14. Fill in the blank with this word: "___ cabinet"

17. Below-ground sanctuary

18. Provincial capital in the Dominican Republic

22. Jazzy Waters

23. The Bible Tells ___'

24. Ticket abbr.

25. Rock's ___ Speedwagon

26. French possessive

27. Suffix with sulf-

28. S.A.S.E., e.g.

32. Work hard

33. Chad's place

34. Workers with 64-Downs, for short

35. Listen: Sp.

36. Name placeholder in govt. records

37. Fill in the blank with this word: ""Ol' Rockin' ___" (bin-mate of the 1957 album "Ford Favorites")"

38. Fill in the blank with this word: ""___-haw!""

41. Cossack chief

42. Steamed dish

43. Submissive type

44. Vital win

45. Recording session need

46. Pinch-hitting great Manny ___

47. Times for showers

52. Not give ___

53. Year in the reign of Edward the Elder

54. Fill in the blank with this word: ""Citizen ___""

56. Philip of 'Kung Fu'

57. Fill in the blank with this word: "___ Men's Health Crisis"

58. Winnebagos, for short

59. Riot-stopping grps.

SOLUTIONS

PUZZLE SOLUTION 1

```
M A R G . E D A T E . C Y S T
K C A R . N O B E T . H A I R
T A K E S T W O T O T A N G O
G R E A T E S T . I O L A N I
. . T A R E . D L I L
. T E N N E S S E E T A N S
C A R T E R . I F S . H I H O
U M W . R E Y . D E M
E M I T . H U N . M A T U R E
. I N D I A N A P O L I S D
. R E R E . A H I S
S A H A R A . A L A B A M A N
W E E W I L L I E W I N K I E
O R M E . D A N S K . E T D S
B I E R . S E T T S . S S E T
```

PUZZLE SOLUTION 2

```
R E A T T A C K S . V O L G A
E X P O S U R E S . E N O R M
O P E R A G O E R . L E W E S
P A R . R U P P . A N A T
E R I . R O S S . C L O S E
N T E S T S . U N I E . T E L
E S S Y . S P E E D I E R
E P I C . V D A Y
T I T I C A C A . R A M P
A Y N . S H N E . I S R A E L
B L E A T . T R E S . I N A
B E R N . U N I V . N A T
E N T O M . A M U S E M E N T
Y O L K S . B E R T L A N C E
S L Y A S . O N E S U I T E R
```

PUZZLE SOLUTION 3

```
G Y P P E D . E A S T L A
A E R A T E D . A Y K R O Y D
S M O T H E R . W E E O N E S
J E J . S M O O C H Y . E S O
E N E S . S W I M S . A T O R
T I C K S . S L O . E S T A B
. S T E E V E . N O S T E P
. L G E . I T A
M A T U R E . F E A R E D
T O T O E . T O R . R T R E V
A B A N . D I R A C . E M M A
M I L . H H M U N R O . A I R
A L O H A O E . C O R N E L L
R E S O R T S . O S S I C L E
A S S I S I . S O C K E T
```

PUZZLE SOLUTION 4

```
F T H S . L O N G B O A T
R A U L . P A N A T E L L A
O G E E . E A V E S D R O P S
M O V E S A H E A D . C F O S
M N O P Q R . N L A K E
. S Y N O D . Q L U P F H
S I S . D E N E B . U S A I D
L E A H . R E R O W . E S S A
Y O G A S . I S A A K . T H Y
S H O R T U . A C C E L
. D U L A C . K N A V E S
I L I A . S C H O O L Y A R D
F A N L E T T E R S . M L I N
I N T E R E S T S . E L E A
M A D E I R A S . N O S H
```

PUZZLE SOLUTION 5

D	O	R	I	C	▪	M	S	G	T	S	▪	S	D	S
A	R	E	N	O	▪	S	M	O	E	R	▪	T	E	C
E	N	D	U	P	I	N	L	A	S	T	▪	A	S	H
D	E	E	P	E	N	▪	▪	W	H	A	T	Y	O	U
A	R	Y	▪	D	E	F	E	R	▪	S	Y	S	T	S
L	Y	E	S	▪	R	E	D	Y	E	▪	C	F	O	S
▪	▪	▪	T	U	T	T	I	▪	S	H	O	O	S	▪
▪	▪	F	U	D	G	E	F	A	C	T	O	R	▪	▪
▪	J	U	D	E	A	▪	I	G	O	O	N	▪	▪	▪
N	A	S	I	▪	S	A	C	E	R	▪	S	A	P	P
O	N	E	O	N	▪	S	E	R	T	A	▪	L	E	A
T	U	L	S	A	O	K	▪	▪	E	D	U	C	T	S
F	A	A	▪	C	L	O	S	E	D	D	O	O	R	S
A	R	G	▪	H	E	U	R	E	▪	O	F	T	E	A
R	Y	E	▪	T	A	T	A	R	▪	N	A	T	L	S

PUZZLE SOLUTION 6

L	A	B	D	A	N	U	M	▪	E	I	D	E	R	S
E	L	E	O	N	O	R	A	▪	L	S	E	V	E	N
H	I	S	S	Y	F	I	T	▪	E	R	R	O	L	L
R	E	S	E	A	U	▪	S	P	C	A	▪	▪	▪	▪
▪	▪	▪	▪	S	H	U	T	T	E	R	B	U	G	▪
▪	I	R	A	I	S	E	▪	O	R	L	E	A	N	S
K	N	E	L	T	▪	A	R	L	O	▪	E	R	I	U
O	C	T	E	T	▪	R	E	E	▪	S	A	K	A	I
V	O	R	T	▪	A	T	O	M	▪	T	R	I	T	T
I	M	I	T	A	T	E	▪	Y	V	O	N	N	E	▪
C	E	M	E	N	T	N	A	I	L	▪	▪	▪	▪	▪
▪	▪	▪	D	U	S	K	▪	A	R	A	I	S	E	▪
G	E	S	U	R	N	▪	R	E	S	O	R	T	E	D
T	O	A	T	E	E	▪	O	M	I	T	T	I	N	G
E	N	N	E	A	D	▪	N	O	C	H	A	N	G	E

PUZZLE SOLUTION 7

K	A	L	E	L	▪	J	O	G	S	▪	W	N	B	A
A	L	A	L	A	▪	A	D	E	E	▪	I	Y	A	M
A	T	S	E	A	▪	M	O	S	H	▪	T	A	M	O
B	E	T	E	L	G	E	U	S	E	▪	H	D	A	Y
A	R	M	▪	A	R	S	L	O	N	G	A	▪	▪	▪
▪	▪	I	S	A	A	C	▪	▪	I	T	A	S	A	▪
A	G	N	I	▪	P	A	D	E	R	E	W	S	K	I
T	A	U	S	▪	P	A	R	L	E	▪	I	K	I	D
W	E	T	B	L	A	N	K	E	T	▪	S	M	T	E
O	L	E	O	S	▪	▪	V	E	T	T	E	▪	▪	▪
▪	▪	O	U	T	E	R	E	A	R	▪	L	I	A	▪
S	L	I	M	▪	I	C	A	N	R	E	L	A	T	E
E	Y	E	B	▪	M	O	T	E	▪	S	I	T	A	R
M	E	S	A	▪	U	L	E	E	▪	S	T	E	N	O
E	S	T	H	▪	R	I	L	E	▪	Y	A	R	D	S

PUZZLE SOLUTION 8

H	U	S	K	I	E	S	T	▪	E	R	E	S	S	O
A	N	H	E	U	S	E	R	▪	T	O	R	E	U	P
U	L	T	I	M	A	T	A	▪	C	A	N	A	P	E
T	A	E	R	▪	S	A	P	P	E	D	▪	L	E	N
E	D	T	▪	T	E	S	T	▪	▪	▪	S	R	S	▪
▪	E	L	E	V	E	▪	Z	A	C	H	▪	K	E	L
▪	▪	T	E	A	S	E	L	▪	A	M	I	G	O	▪
R	E	A	R	G	U	E	▪	M	A	G	I	N	O	T
A	F	L	E	A	▪	R	U	E	F	U	L	▪	▪	▪
M	F	A	▪	N	D	A	K	▪	S	E	T	A	E	▪
S	L	R	▪	U	P	R	I	▪	▪	▪	R	L	S	▪
H	U	M	▪	B	L	E	A	R	S	▪	P	E	L	E
O	V	I	S	A	C	▪	I	V	E	G	O	T	I	T
M	I	S	S	M	E	▪	N	I	C	E	S	H	O	T
E	A	T	S	A	T	▪	E	N	T	R	E	A	T	S

PUZZLE SOLUTION 9

```
A F F I L I A T E ■ S H O E B
G O E S O F F O N ■ D E W A R
E R N I E F O R D ■ S I N G A
I T D ■ ■ Y U M A ■ ■ N S E C
S W E E T ■ L E N ■ A Z U R E
M O R T E M ■ N G O R ■ P E R
■ ■ ■ A D E ■ T E N T ■ T S O
■ A S L E E P ■ R E B O O T ■
E N L ■ U C A L ■ B O L ■ ■ ■
S N E ■ M E S A ■ C O D D L E
C M D R S ■ T Y S ■ K H E A D
H A D A ■ ■ R I C H ■ C D E
E R I N S ■ A T R O C I O U S
W I N D Y ■ M O O N R O C K S
S E G A L ■ I N D I A N T E A
```

PUZZLE SOLUTION 10

```
L S A T S ■ ■ S A S H E S ■
O U N C E S ■ O T R A N T O
W R A I T H S ■ T R A N C E S
M G T ■ S M U S H E S ■ A N A
A E O N ■ O S S I E ■ S M O G
R O L E N ■ A E S ■ L A P S E
K N E E B O N E ■ C U R S E S
■ ■ D C O N ■ P I T A ■ ■ ■
E I G H T S ■ A I R S P A C E
S C H A V ■ E S Q ■ K E V A N
S I A M ■ P N S U P ■ S I G S
A C S ■ K E R N E L S ■ A I T
I L T E M P O ■ T O O K T E A
S E L L S T O ■ D R I E S T
■ D Y E P O T ■ ■ T A S T E
```

PUZZLE SOLUTION 11

```
I B E T ■ A D I S H ■ C P O S
T C H R ■ L A R C H ■ O H B E
I C E E ■ G R O S S ■ C L O R
N I L E ■ E I N ■ W O O L F
■ ■ T U R N O N A N A X I S
G E O R G ■ S U E R S ■ ■ ■
A N N U L S ■ T B A ■ S P R S
I T S N I C E T O B E H E R E
L E E K ■ H R H ■ S L I P U P
■ ■ A M I E S ■ O R A N T
P U T A S O C K I N I T ■
S N A T H ■ I M O ■ T A O S
S H N E ■ M I N O R ■ A P S E
T I T I ■ M A K O S ■ I I I I
S P E T ■ I T S M E ■ L A P S
```

PUZZLE SOLUTION 12

```
S A N T ■ N E W T ■ A H O S T
Q U A I ■ I T S O ■ N A D I A
F R O M S T E M T O S T E R N
T A H ■ M E R S ■ F W A R E
■ ■ P O R N ■ B A E R ■ ■
P R O P O S E M A R R I A G E
C O O P T ■ A T T S ■ M E L
L O N S ■ M S S ■ I I I I
A N A ■ A V E O ■ F E T C H
B E S T T H I N G F O R Y O U
■ ■ A T O R ■ O R S E ■
S E R I N ■ I B I S ■ T B A
N O M O R E I L O V E Y O U S
A L A T E ■ T I T O ■ A P R I
L E G S D ■ N A S L ■ O U S E
```

PUZZLE SOLUTION 13

```
A A B B . A M S O . B P E E N
C L I O . D A Y S . R E B B E
H O O F . H U N T . E R R O N
S E L F P E R C E P T I O N .
. N R A . O A T S .
S P O U S E . O P U S . R O B
C O R F U . G I A S . D E B S
T I T O P U E N T E T E P E E
E S O S . G O G H . R E E S E
R E N . A L L O . S U R G E D
. B R I O . M T N .
. T H A T S G R E E K T O M E
W H A R F . I A L L . H D A Y
L E K E U . S E E M . E E E E
E Y E S L . T R E O . A S O B
```

PUZZLE SOLUTION 14

```
A N T S . O N A N . U P F O R
B O A C . R O T I . N A R D O
B U S H V G O R E . B R A E S
A N E M I A . I N S E C U R E
. O N N A . T T L .
. T H E Y D O P E N D A N T S
T H A . L I N E . I L I U M
O R T H . E E L E D . G N M A
N O H I T . E N O L . T M S
I W A K E U P G R O U C H Y .
. R H O . Y R L Y .
S P I T E F U L . B U R S T S
A R S O N . T A K E S A S I P
S O N I C . E M I L . N E N A
H A O L E . R E A L . O T T S
```

PUZZLE SOLUTION 15

```
C A R P . M P E G . B A D A T
C S I S . A H S O . A L U M S
C A C A O T R E E . S K I E R
P H E N O M . S T U S .
. I S D N A O . H S I T T E R
. Q A N N E . E S S E N E
A M E S S . E S T A T E C A R
N E R . A R T I S . T R U
J A M E S C A A N . A D A M N
O N A B E T . R E B E C .
U S S W A S P . D S T U D S
. G O L D . E N T E R O
F I C H U . E L E V A T I O N
A N E E L . B I N E . E T O S
A F O F L . E L A N . R E T E
```

PUZZLE SOLUTION 16

```
H E R S . A N T . I N U S E
E R S E . C O O N . S O N A R
P A V E . D I N O . A R D O R
T O P S E C R E T C O D E .
O F D A Y . A M O . R S A
. W E L L R E G R A D E D
S L E . E D M . N E R O L I
D E X T R A L . H A I R N E T
L A T H E D . A U T . E S S
T H R E E S I S T E R S .
G S A . T S O . Y E S O R
. F L Y O F F T H E W A L L
A B I E S . O N I T . A B E E
U L N A E . R O T T . G R I S
D O E R R . W I M . E E N S
```

PUZZLE SOLUTION 17

M	T	H	S		O	B	E	Y		T	A	F	T	S
A	H	O	T		R	I	R	E		R	U	D	E	R
A	A	A	A		G	L	E	S		E	R	I	N	S
M	E	X	I	C	A	L	I	M	E	X	I	C	O	
		R	U	N	G			S	E	C				
C	L	O	S	E	S		T	I	T	S		S	P	R
R	O	S	T	I		L	I	A	O		M	W	A	H
A	T	T	E	N	T	I	O	N	P	L	E	A	S	E
S	T	E	P		R	E	G	D		U	N	S	E	T
H	E	R		S	E	D	A		U	P	S	H	O	T
		F	P	A			I	N	S	S				
	B	R	E	A	D	A	N	D	B	U	T	T	E	R
I	L	I	A	C		A	L	S	O		O	R	E	N
S	A	N	T	E		B	E	A	R		R	A	R	A
L	E	G	S	D		B	R	Y	N		E	N	O	S

PUZZLE SOLUTION 18

A	P	A	S	S		S	O	O	T		V	E	R	Y
T	H	R	O	E		H	W	F	E		I	L	I	E
B	A	T	H	W		M	O	A	N		N	E	I	N
	R	A	I	N	B	O	W	G	O	D	D	E	S	S
		O	I	O			E	R	E	I				
A	M	A		N	I	E	R		S	L	E	E	P	Y
S	E	T	I		T	R	E	N		O	S	T	I	A
S	U	R	G	E	O	N	S	G	E	N	E	R	A	L
A	S	A	N	A		S	E	E	P		L	E	S	E
D	E	P	O	N	E		E	D	H	S		S	T	S
		R	O	P	Y			O	A	K				
N	A	T	A	L	I	E	P	O	R	T	M	A	N	
E	G	A	N		L	S	A	T		O	A	S	I	S
I	N	S	C		O	W	I	E		R	R	S	T	A
L	I	M	E		G	E	N	A		I	T	T	E	R

PUZZLE SOLUTION 19

S	E	L	L		H	E	R	E		C	L	E	A
M	E	T	A		A	L	E	F		T	H	I	R	D
S	K	Y	D	I	V	I	N	G		R	A	M	I	E
H	A	R	D	C	A	S	E		P	A	L	A	T	E
		A	S	I		T	O	N	E	S	U	P		
W	O	N	T	H	U	R	T	A	B	I	T		
S	P	A	H	N		E	R	O	O		G	P	O
M	A	R	X		F	E	D	E	X		D	I	A	N
S	L	Y		K	A	L	I		D	O	L	C	E
		P	I	C	K	U	P	T	H	E	T	A	B
T	R	A	I	N	E	E		L	A	A			
R	A	T	E	D	R		O	U	T	B	U	R	S	T
I	R	A	T	E		V	E	N	T	I	L	A	T	E
P	I	L	A	R		E	N	K	E		L	N	E	S
E	N	L	S		A	S	S	D		A	D	M	S

PUZZLE SOLUTION 20

T	I	S	C	H		N	O	R	I		S	L	A	Y
I	O	W	A	Y		C	O	E	N		T	E	M	A
E	N	A	R	M		A	R	T	S	T	U	D	I	O
D	E	M	O	N	R	A	T	I	O	N	S			
	S	I	L	A	S		P	L	U		O	L	A	
		I	L	I	C	H		E	T	I	M	E	S
S	A	I	N		D	E	L	E		S	R	E	P	A
A	C	T	E		E	R	A	O	F		R	R	U	N
I	C	O	N	O		E	N	N	A		E	S	S	A
D	E	F	O	R	M		D	S	T	A	R			
A	L	F		B	U	C		S	U	S	I	E		
		R	I	T	A	C	O	O	L	I	D	G	E
I	M	S	O	T	I	R	E	D		A	B	E	E	T
S	L	O	W		N	O	L	I		I	L	A	S	T
O	K	I	E		G	M	A	C		T	E	S	T	A

PUZZLE SOLUTION 21

```
A W I L L █ F H U P █ M B A S
C A C A O █ L A C E █ I L I E
C R A Z Y Q U I L T █ G E S T
T E L E O S T █ A S P H A L T
█ L N E S █ E T T E S
S G T M A J █ M E R R Y
A Q H S █ V A L U E █ Z C Q
C H I N E S E C H E C K E R S
S H S █ M B E K I █ E R I Q
█ G O A P E █ N I N O N S
I T S A N █ D A S H
S U N R O O F █ D E E P E N D
O N E I █ C L U B C A R T O N
L A R S █ T A L I █ R A N D A
E S T H █ A N T Z █ D M A S S
```

PUZZLE SOLUTION 22

```
C H O M P █ T R A S H
W O O D I E R █ S H O T P U T
A R G O N N E █ P E D A L E R
K Y A N I T E █ H O S T I L E
E Z R A █ V E E S █ I C E S
N A T T Y █ E U R O █ M E S S
S H A L L █ R I P P E R S
E A R A C H E
G R A M M E S █ Y E M E N
O R A L █ I B I S █ R E V U P
C A G E █ N A A N █ M E R E
E N T R E A T █ I M M E N S E
A G I T A T E █ P O I N T E R
N E M E S E S █ E A S T E R S
R E D E D █ S T O R Y
```

PUZZLE SOLUTION 23

```
D I A Z █ I P S A █ K S T A R
E D G E █ G O T O █ E C O L E
B I C A M E R A L █ A R R I D
S O Y L A T T E █ E N U R E S
█ R N A █ O X U F
S P O R T O █ I N F █ F A H D
T O S A Y █ O N E B █ M A O
I N S U R R E C T I O N A R Y
C D I █ I S A O █ P U T T O
H S E A █ A T S █ M I L I E U
█ R Y N E █ E A N
E E L P O T █ I N N E R A C E
A V I E W █ F O O D S T A M P
D I E L L █ I N R E █ E R O O
S E U S S █ N O E L █ E E N S
```

PUZZLE SOLUTION 24

```
E R N S █ J S U P █ S A I L S
P E A U █ O H N E █ C P L U S
H A L F P R I C E █ R I E N S
A S A █ A J A █ L E A R N S
█ R S A █ E R L E S
B L O A T █ P L E O N A S M S
A Y E N █ S U E T Y █ W E A N
M R S D █ E L C I D █ A T M O
B I T O █ E S T E S █ P H E W
I C E M A K E R S █ P I S T E
█ N O T S O █ S T A
T E S O R O █ R A U █ A V S
T R A I T █ F O R H I T L E R
O D E S A █ A R E E █ H A N A
P A S E S █ A D D L █ O S A S
```

PUZZLE SOLUTION 25

R	S	V		L	E	G	T	O		H	I	S	T	O
E	M	I		C	R	E	A	M		D	O	T	E	R
D	I	V	I	D	E	D	C	A	P	I	T	A	L	S
S	L	I	T				T	H	E	R	A	M		
K	L	A	T	S	C	H		A	D	J		P	I	T
Y	A	N		T	U	E	S		E	B	B	E	T	S
		K	A	R	S	T	S			G	D	A	Y	
	V	O	W	E	L	S	I	N	O	R	D	E	R	
T	I	N	A		E	R	I	C	I	V				
D	O	T	I	N	G		S	P	A	R		R	H	E
P	L	O		E	A	P		S	N	E	A	K	E	D
		G	E	A	R	E	D			B	E	Y	E	
T	W	E	N	T	Y	N	I	N	E	P	A	L	M	S
P	E	N	N	E		T	C	E	L	L		L	O	S
S	A	Y	E	R		A	E	G	I	S		Y	M	A

PUZZLE SOLUTION 26

S	A	N	T		E	D	I	F	Y		O	P	T	S
C	L	A	W		M	U	S	E	E		N	I	A	S
A	T	M	O		T	O	O	T	H	P	A	S	T	E
R	H	U	M	B		D	N	A		I	T	H	E	E
F	O	R	O	N	C	E		L	I	Q				
			M	A	I	N	S		N	U	D	I	E	S
C	H	E	M	I	C	A	L		R	E	R	A	C	K
C	O	S	I		H	L	I	N	E		A	T	O	I
U	N	S	E	A	L		M	E	A	N	W	E	L	L
P	I	U	S	I	I	I		Y	W	C	A	S		
		M	D	C		W	H	I	T	N	E	Y		
U	P	S	E	T		O	S	A		R	A	I	D	S
X	R	A	Y	O	R	I	N	G	S		R	O	N	A
O	I	S	E		B	L	E	E	P		E	B	A	Y
R	E	E	S		I	S	E	R	E		S	E	S	E

PUZZLE SOLUTION 27

M	E	S	A		P	O	P	E		A	B	A	T	E
A	M	P	S		A	P	E	X		R	O	W	A	N
L	E	A	S		R	A	R	E		R	O	O	T	S
I	N	D	E	S	T	R	U	C	T	I	B	L	E	
	D	E	S	A	L	T		I	V	Y				
		S	L	Y		P	O	R	E		A	P	T	
I	D	I	O	M		B	R	I	E		I	C	A	O
S	U	N	R	I	S	E	I	N	D	U	S	T	R	Y
L	O	S	S		C	A	S	K		L	O	S	E	S
E	S	T		D	O	R	M		A	T	M			
		O	A	R			S	U	R	E	T	Y		
	D	I	S	I	N	C	L	I	N	A	T	I	O	N
F	I	N	I	S		H	U	N	T		R	A	G	E
E	M	C	E	E		A	S	C	I		I	R	I	S
Y	E	A	R	S		T	H	E	E		C	A	S	T

PUZZLE SOLUTION 28

B	L	A	H		G	R	A	P	H		L	O	U	R
A	I	N	U		R	A	N	E	E		E	U	R	O
D	O	U	B	L	E	S	T	A	N	D	A	R	D	S
E	N	S		Y	E	T	I		E	N	S	U	E	
		A	N	N	A		W	E	F	T				
D	E	L	U	X	E		L	I	M	A		S	E	A
O	X	E	N		T	O	N	I	C		P	A	L	
W	E	A	T	H	E	R	F	O	R	E	C	A	S	T
S	A	C		U	N	I	T	S		O	R	E	O	
E	T	H		R	I	M	Y		B	E	V	E	L	S
		O	D	D	S		C	A	P	E				
S	P	O	O	L		R	A	N	I		G	E	E	
A	L	I	M	E	N	T	A	R	Y	C	A	N	A	L
Y	E	L	P		A	O	R	T	A		T	A	S	K
S	A	S	H		P	R	E	E	N		E	W	E	S

PUZZLE SOLUTION 29

H	B	O	M	B		A	B	A	F	T		P	A	P
E	L	G	A	R		C	O	L	A	S		R	I	O
W	O	R	D	O	F	M	O	U	T	H		O	D	E
S	T	E	R	I	L	E		M	E	I	O	S	E	S
			A	L	E			D	R	O	P	S	Y	
A	L	B	S		N	O	N	O		T	H	E		
P	A	L			S	C	A	L	P		E	R	G	O
S	M	U	G		E	C	O	L	I		D	I	L	L
E	P	E	E		S	U	M	A	C			T	A	D
	R	N	A		R	I	S	C		G	Y	M	S	
S	C	I	O	N	S			O	B	I				
T	O	B	A	C	C	O		A	L	A	B	A	M	A
O	R	B		H	O	T	C	R	O	S	S	B	U	N
A	G	O		O	P	I	U	M		R	O	U	S	T
T	I	N		R	E	S	T	S		A	N	T	E	S

PUZZLE SOLUTION 30

T	A	R	P		P	S	S	T		S	N	O	B
O	D	O	R		I	C	H	O	R		H	I	R	E
D	A	T	E		N	A	I	R	A		U	S	E	R
D	R	A	C	O		G	R	A	D	A	T	I	O	N
		R	E	N	D		R	H	I	N	O		
S	T	Y	P	T	I	C		S	I	T	U	P	S
P	E	C	T	O	R	A	L		S	T	R	O	P
U	R	L		T	R	I	T	E		O	N	O	
D	R	U	I	D		P	H	Y	S	I	C	A	L
	A	B	R	O	A	D		Y	A	M	M	E	R	S
	A	T	S	E	A		S	U	P	S		
D	E	A	T	H	T	O	L	L		T	A	S	T	E
A	C	R	E		I	D	I	O	T		S	I	O	N
S	H	U	L		R	A	B	B	I		T	O	L	D
H	O	M	Y		R	I	S	C		O	N	U	S

PUZZLE SOLUTION 31

A	K	A		M	A	A	S		U	S	U	R	Y
C	I	N	C		O	P	T	O		T	H	R	O	E
A	R	T	H	U	R	I	A	N		A	T	L	S	T
S	O	R	A	R	E		D	O	A		E	S	A	I
E	V	E	L	K	S			N	A	T			
	L	E	O	P	O	L	D	B	L	O	O	M	
S	I	S	A	L		O	Z	A	R	K		N	M	I
I	T	C	H		K	L	I	N	E		S	A	A	R
K	L	U		E	R	I	C	A		A	U	T	R	E
A	L	P	I	N	E	S	K	I	I	N	G		
	K	G	B			C	O	G	N	A	C		
C	L	I	O		S	O	B		A	N	E	C	D	E
R	E	D	I	D		H	A	D	R	O	S	A	U	R
A	L	I	K	E		T	H	R	U		T	A	L	E
M	Y	G	O	D		O	N	U	S		S	T	S	

PUZZLE SOLUTION 32

D	C	A		C	O	R	F	U		B	A	R	M	S
I	R	T		U	R	A	L	S		R	I	E	N	S
N	O	T	G	I	V	E	A	H	O	O	T	F	O	R
E	S	E	L			P	E	N	N				
A	S	S	I	G	N	S		R	E	T	A	I	L	S
T	E	T		R	A	L	F		H	E	F	N	E	R
		C	U	M	U	L	I			O	C	A	S	
	V	A	N	B	U	R	E	N	S	W	R	E	N	
L	O	U	D		S	A	T	I	R	E			
O	L	E	O	L	E		S	R	T	A		M	C	M
N	E	L	S	O	N	S		A	S	P	H	A	L	T
	G	A	O	L				E	P	E	E		
B	R	I	T	I	S	H	R	A	I	N	C	O	A	T
A	N	N	A	N		I	O	N	I	C		U	R	N
L	A	T	H	S		O	N	A	I	R		T	S	A

PUZZLE SOLUTION 33

PUZZLE SOLUTION 34

PUZZLE SOLUTION 35

PUZZLE SOLUTION 36

PUZZLE SOLUTION 37

```
D O V E B A R . F T D . C P L
A T H L E T E . I B E . D L E
H O F D A R K N E S S . L A P
. . . T Y N E . I M I N E
T H U M B . O H I O R I V E R
R A N P A S T . A S E A .
A I R A C E . D S M . S A C
S T E A K M E D I U M W E L L
H I P . P O T . N E A R T O
. E V E N . B D A L T O N
R E H E A R S A L . L E A S E
S H O W N . C A P T .
I U M . I N N U M E R A B L E
D D E . T S E . E P A L A B S
E S S . Y A Y . R A Y K R O C
```

PUZZLE SOLUTION 38

```
V I E R . H O E D . I N R E M
I T S A . O R L E . L E O X I
C O O T . T A U S . T R A P S
I N S T I L L L I F E . D O T
. A H I . C O M . R S A
Y O K N A P A T A W P H A
E P A . D S O S . L O W C U T
D A T A . K A R . Y I P S
O L E F I N . R O P O . N R C
. A T H E N S G E O R G I A
I S A . E E E . E M O
N H L . A R C H I P E L A G O
M A L A R . T A H E . L R O N
A M I T Y . A M I R . U M B O
N U E V A . R E T S . P A S S
```

PUZZLE SOLUTION 39

```
H A E C . D A C H E . B M A J
T A R T . E L R O Y . R E N A
T H I N G P I E C E . A A C Y
P S U . E L E E . G L I D E S
. P N O N . B L E D
S T A R R Y . S U A V E S T
P I Q U E . H A R S . D H O W
A N U N . I A M B S . M A R E
D A A E . L B O S . H A N S A
S E D A L I A . L O N D O N
. A D A B . T O L E
L E T N O T . A B O Y . M O B
O M E I . E A S I N G I N T O
P A N S . E M I L E . T E E S
E G G H . E A R L Y . T M A N
```

PUZZLE SOLUTION 40

```
O N T A P . B L A S T . P S S
R O S T I . U I N T A . A L T
H A R M F U L L L I G H T R A Y
E M I . F S L I C . R O L L
. S O L U S . E M M Y L O U
T E T H E R E D . T O T E M S
Z W E I . Y Y B . V S O
U A R . E A S . M T N
. S A K . S E M . S A Y A
B I C O R N . E X A M I N E E
E G A L I T E . D C E L L
S O M E . N O R A S . I B A
S T E A K M E D I U M W E L L
E T A . K I R O V . E L S O L
S A T . K R O N E . R E T C H
```

PUZZLE SOLUTION 41

PUZZLE SOLUTION 42

PUZZLE SOLUTION 43

PUZZLE SOLUTION 44

Solution 41
```
C T R L . F D I C . . D A N L
O H I O . U P L I T . T N P K
O R M E . E V I T A . E A R P
P O I S O N W A R N I N G .
T E E S H T . O N O . R I S
S S R . N E B S . I N C A N S
. B O S T O N C O M M O N
O G P U . W Y O . L S T S
H A R R Y H O U D I N I
H O O P O E . Z E A L . O S T
I L B . S R A . S A I N T E
. A L T E R E D S T A T E S
A S T A . T E D I U . G A P S
F L E W . O N E A M . O P I A
T O D S . A L L E . S E N S
```

Solution 42
```
T O N E . S A H L . T A L U S
O M A N . T H E E . E L I S A
O N N O . O O R T . C A F F E
K I C K U P Y O U R H E E L S
H A Y I N G . N P I N
. F O U R . C O N M A N
O B O L I . G Y R A . A A B A
F R A C T A L . A N S P A C H
I N S T . M I T T . U S R D A
T O T O I V . A S A D
. N E A P . S O B E R S
W H A T S T H E B I G D E A L
R I C K I . O M I T . O N T O
A T H O S . S A T I . O I S E
P S S S T . T N T S . R E O S
```

Solution 43
```
P I T H . O N U P . . G E A R
A M O O . R I C E D . A L T A
V I S I T I T S W E B S I T E
E T H . O N E D . V A L E N S
. M M D . D I D I
B A S E B A L L B A T T E R S
O P A L S . S E A T O . R O T
Y O N D . P E N C E . J I B E
E S D . N O V A K . O U T O N
R E P O S S E S S E D A U T O
. R Y A N . R E N
A L M O N D . E E R O . B O A
M E X I C A N S T A N D O F F
O V I D . S A T A N . K O F C
R I V E . A D H D . A N S A
```

Solution 44
```
C A T S . R A P T . S H A M U
W L O O . I D E O . N E W E R
T E R I . C A P O . O C E A N
S C I S S O R S K I C K S .
. E P T . I T S A . C S T
D E F E A T S . O U T G R E W
O V I . H A R M . I A M I
S E V E N S I X T Y S E V E N
I L E R . C O E N . E L E
N Y G I A N T . P L A I N E R
G N U . I A I M . L I D
. I N T E R S T E L L A R D
C O N T I . A T A D . E S S I
E R E W E . N I N A . R C T S
Y E A T S . A C A T . S H U T
```

PUZZLE SOLUTION 45

A	M	P	S		I	L	E	A		M	I	N	D	S
R	U	T	H		N	E	B	R		C	R	I	E	S
A	S	B	A	D	A	S	B	A	D	C	A	N	B	E
B	L	O	W	O	N		W	I	L	E	E			
L	I	A	N	G		P	S	A	S		H	M	O	
E	N	T		O	Z	I	C	K		A	F	O	F	L
			D	O	E	R	R		O	D	E	L	A	Y
	C	A	R	D	I	N	A	L	R	U	L	E	S	
M	I	D	G	E	T		W	E	I	L	L			
A	D	E	E	R		A	L	A	N	A		O	R	L
Y	E	N		L	T	Y	R		T	E	H	E	E	
	A	D	A	I	R		M	E	T	A	L	S		
M	O	U	N	T	A	I	N	O	U	S	A	R	E	A
B	R	E	A	M		U	N	O	S		G	A	N	G
B	I	R	S	E		M	E	N	D		E	S	T	E

PUZZLE SOLUTION 46

W	A	S	H	O		B	B	S		N	A	A	C	P
O	R	T	E	A		A	R	A		E	L	I	H	U
A	L	E	C	S		R	A	I	N	W	A	T	E	R
D	O	N	T	P	L	A	Y	D	O	M	E			
	I	R	E		H	T	O		O	I	E			
S	P	A	C	E	O	F	D	I	A	M	O	N	D	S
E	E	R		E	X	I	M		S	L	E	E	T	
N	A	O	H		I	B	A	R	S		E	T	A	H
T	H	U	M	B		R	E	E	D		O	T	E	
R	E	N	O	R	O	C	K	C	O	R	O	N	E	R
A	N	D		E	P	I		U	S	O				
	T	R	A	C	E	E	L	E	M	E	N	T		
S	H	E	L	F	H	E	L	P		U	P	F	O	R
P	I	S	M	O		L	O	L		S	A	L	L	E
R	E	L	A	X		Y	I	N		S	H	A	L	T

PUZZLE SOLUTION 47

E	R	T	E		A	A	A	A		H	A	K	I	M
D	E	A	N	S	L	I	S	T		E	G	A	L	S
T	I	L	T	A	T	W	I	N	D	M	I	L	L	S
V	M	I		Y	M	A		A	S	T				
	S	A	C	H	A		D	E	N	I	A	B	L	E
	A	I	N		N	U	N	N		R	E	A		
F	A	I	N	T		L	I	R	E		O	U	I	S
O	R	F	E	O	E	D	E	U	R	I	D	I	C	E
U	T	E	R		C	O	P	S		M	O	N	A	D
N	O	A		A	L	P	E		C	M	R			
D	O	R	A	M	A	A	R		R	E	S	T	S	
	L	O	I		E	A	R		B	O	S			
P	E	P	P	E	R	M	I	N	T	S	T	I	C	K
S	T	I	E	B		D	I	N	E	E	A	R	L	Y
S	A	M	S	A		S	N	E	D		I	D	E	E

PUZZLE SOLUTION 48

B	A	S	E		S	M	A	R	M		S	O	L	I
E	F	A	X		E	C	L	A	T		P	S	A	T
A	L	I	C	I	A	K	E	Y	S		U	S	I	A
R	E	D	E	L	M		A	S	T	E	R	I	C	K
		L	E	Y	E		R	N	A	S	E			
L	H	A	S	A		V	E	X	E	R	S			
A	D	I	A		D	A	M	A	T	O		I	R	R
M	A	R	T	H	A	D	A	N	D	R	I	D	G	E
B	Y	S		T	H	E	T	A	S		L	E	W	D
	F	E	S	S	E	D		F	L	O	J	O		
I	N	H	I	S		U	R	D	U					
P	O	I	N	T	S	T	O		A	I	S	L	E	D
O	L	L	A		S	I	M	U	L	C	A	S	T	S
D	I	L	L		G	R	A	S	P		G	A	T	O
S	E	S	E		T	O	R	C	H		E	T	E	S

PUZZLE SOLUTION 49

O	T	O	H		R	S	T	U		M	T	A	P	O
N	O	N	U		E	K	E	S		E	I	D	E	R
E	N	E	M	Y	M	I	N	E		S	M	I	L	E
H	O	O	P	O	E			B	A	S	E	M	E	N
		H	U	N			H	Y	D	R	O			
F	I	R	S	T	D	A	Y		U	S	U	R	P	S
O	N	A		O	S	C	U	L	E		T	I	R	O
G	U	I	D	O		H	N	S		W	S	T	O	N
U	R	S	I		M	T	D	A	N	A		T	M	I
P	E	A	R	C	E		A	T	C	H	I	S	O	N
		T	A	M	M	I		W	O	N				
O	I	L	Y	G	O	O		Y	O	G	U	R	T	
B	A	I	R	N		T	I	D	E	S	O	V	E	R
A	N	T	A	E		E	C	H	T		T	E	T	O
N	A	S	T	Y		T	E	S	H		S	A	D	D

PUZZLE SOLUTION 50

I	N	S	E		S	M	S	H		S	M	E	L	L
N	O	W	A		L	I	E	U		E	A	R	T	O
S	T	E	S		A	R	A	M		L	I	B	R	A
T	I	P	T	O	P	O	N	E	I	L	L			
R	E	T	O	L	D		S	R	I		R	I	F	F
		D	Y	A	D		I	N	C	O	L	O	R	
S	B	A		M	S	R	P		R	O	L	L	E	
T	E	L	E	P	H	O	N	E	N	U	M	B	E	R
A	L	T	A	I		C	R	O	M		E	Y	E	
F	I	E	R	C	E	R		S	T	B	D			
F	E	R	N		P	E	A		L	L	O	Y	D	S
		I	T	I	S	T	H	E	Y	E	A	S	T	
A	W	I	N	G		I	T	E	A		S	L	O	E
B	G	W	G	L		N	I	M	S		I	T	U	P
A	N	O	S	E		S	C	A	T		N	A	P	S

PUZZLE SOLUTION 51

U	S	C	G		L	O	A	D	S		M	E	E	R
S	H	O	E		O	R	F	E	O		C	A	M	P
E	A	R	N		S	T	A	R	B		C	R	I	T
F	L	O	O	R	T	O	C	E	I	L	I	N	G	
U	L	N	A	E		E	G	G	B		I	R	E	
L	I	A		L	I	I		J	O	N	A	S		
		D	I	P	P	E	R	S		S	G	N	Q	
	P	E	R	C	U	S	S	I	O	N	I	S	T	
M	L	B	A		T	E	L	E	S	I	S			
D	E	O	X	Y			L	O	C		E	S	T	
T	A	N		T	A	T	A		H	I	N	T	S	
	S	I	N	K	S	T	O	A	N	E	W	L	O	W
T	A	T	E		R	I	L	L	E		I	A	M	A
M	N	E	M		E	M	E	A	T		S	C	A	N
I	T	S	O		D	E	R	N	S		H	E	S	A

PUZZLE SOLUTION 52

B	A	T	T		H	L	A	N	D		U	G	L	I
O	T	H	O		E	E	L	E	R		E	L	A	L
W	O	R	M		T	A	E	B	O		C	E	R	O
	P	O	W	D	E	R	R	O	O	M	K	E	G	S
	O	U	R	N			P	R	E	S	E	T		
A	D	O	L	F	O		A	L	I	A	R			
B	L	U	F	F		J	E	R	E	Z		U	M	W
R	E	T	E		S	I	R	O	R		X	R	A	Y
I	G	A		A	L	L	I	N		E	R	I	C	S
	E	C	O	L	E		C	E	A	S	E	S		
A	M	E	N	D	E			A	H	O	Y			
P	O	L	I	C	E	O	F	F	I	C	E	R	S	
L	O	G	S		Y	O	G	I	C		Y	E	E	S
U	R	A	L		E	L	O	R	O		E	L	M	S
S	E	R	E		D	O	Z	E	S		S	O	S	A

PUZZLE SOLUTION 53

T	A	B	U	L	A	R	A	S	A	■	M	A	C	S
A	G	I	T	A	T	I	O	N	S	■	A	G	R	I
S	E	N	A	T	E	C	L	O	C	K	R	O	O	M
M	E	S	H	■	S	K	E	W	■	N	Y	A	C	K
■	■	■	S	T	I	R	S	■	A	I	L	E	S	
■	O	C	H	O	S	■	■	E	P	I	■			
C	H	A	R	M	■	Y	M	I	R	■	E	B	W	
L	I	K	E	A	B	U	M	P	O	N	A	L	O	G
E	O	E	■	■	I	M	S	O	■	T	E	M	P	T
■	■	L	I	N	■	■	S	S	S	S	S			
E	O	L	I	C	■	H	A	B	I	B	■			
A	S	O	N	E	■	A	L	O	F	■	A	N	E	S
T	R	A	N	Q	U	I	L	I	T	Y	B	A	S	E
M	I	C	E	■	S	T	A	T	E	M	E	N	T	S
E	C	H	T	■	L	I	T	E	R	A	T	U	R	E

PUZZLE SOLUTION 54

D	I	C	T	■	A	D	A	M	S	■	C	N	N	
A	B	O	O	■	T	I	A	M	O	■	R	A	E	S
H	I	S	O	L	D	S	E	L	F	■	O	P	R	Y
■	■	T	I	A	S	■	T	Y	C	O	O	N		
R	E	C	H	E	W	■	S	S	S	A	I	L		
O	R	I	G	I	N	■	E	L	O	N	■	E	T	A
A	S	T	E	N	■	E	L	E	A	N	■	O	O	N
D	A	I	L	■	S	L	E	E	P	■	T	N	P	K
E	T	E	■	T	H	I	C	K	■	R	A	I	T	A
O	Z	S	■	A	U	S	T	■	D	E	N	I	E	R
■	O	A	K	T	A	G	■	A	N	G	I	N	A	
B	I	N	G	E	D	■	A	N	T	I	■			
I	N	T	O	■	O	P	P	O	S	A	B	L	E	T
T	A	H	R	■	W	A	C	K	O	■	L	I	K	E
■	S	E	A	■	N	O	T	I	N	■	E	Z	E	K

PUZZLE SOLUTION 55

A	L	M	S	■	P	L	O	W	S	H	A	R	E	S
W	A	U	L	■	L	O	C	A	L	I	T	I	E	S
H	I	T	O	V	E	R	T	H	E	F	E	N	C	E
I	N	T	■	A	N	D	A	■	D	I	O	■		
R	I	O	■	L	A	E	■	A	S	S	U	M	E	D
L	E	N	D	L	■	D	S	T	■	T	I	R	E	
■	■	C	R	E	T	■	A	T	O	E	■	A	D	R
■	T	H	E	J	O	E	W	A	S	O	N	M	E	
Y	A	O	■	O	M	I	T	■	O	M	R	I		
A	M	P	M	■	R	O	O	■	E	C	A	N	S	
D	I	S	E	A	S	E	■	B	A	L	■	D	O	A
■	S	L	A	■	M	T	G	E	■	V	O	L		
T	O	P	S	O	F	M	O	U	N	T	A	I	N	S
O	D	E	R	N	E	I	S	S	E	■	N	C	A	A
N	O	R	S	E	R	A	C	E	S	■	C	E	N	S

PUZZLE SOLUTION 56

■	E	L	E	C	T	S	■	P	O	S	A	D	A	■
A	T	O	T	H	A	T	■	R	L	S	T	I	N	E
G	A	U	D	I	E	R	■	O	S	I	E	R	E	D
E	L	I	■	C	L	A	M	P	O	N	■	E	M	I
N	I	S	I	■	S	I	T	O	N	■	N	C	O	S
T	A	X	C	O	■	N	E	S	■	L	E	T	N	O
S	E	V	E	N	S	E	V	E	N	S	E	V	E	N
■	A	C	A	D	E	M	Y	A	D	■				
T	A	R	G	E	T	P	R	A	C	T	I	C	E	S
I	S	E	E	A	■	E	E	R	■	S	E	H	N	A
N	A	P	S	■	B	A	S	R	A	■	R	O	L	L
O	R	I	■	T	A	C	T	I	L	E	■	R	A	U
R	U	N	T	I	S	H	■	A	T	T	R	A	C	T
E	L	E	V	A	T	E	■	G	A	L	I	L	E	E
■	E	R	A	S	E	S	■	E	R	A	S	E	D	

PUZZLE SOLUTION 57

```
A C E R   T A R O   S E T O F
L A R C   A B A S   A G O R A
T R A V E L S S O S L O W L Y
A N O R A K   P N C   I S E E
R E F   U S X   G O A D
        S H A R   P L E A S E
A R M   O N E A   L A B O R
T H E C O W A R D L Y L I O N
E Y E R S   X O R O   G T I
E S T E E M   W A S I
    A E A N   W T O   S H A
C A T T   M E I   A L E A S T
T R A I N E D A S S A S S I N
R E E V E   D I K E   O H N O
S E R E R   A M D A   S A G S
```

PUZZLE SOLUTION 58

```
T O D D   M V I I   R E B A S
O D A Y   C O G S   F R A H T
M O D E   G L E S   D E R M A
B R O W B E A T E R S   R A G
    O V E R A L L   F E D S
S C R O D S   E A R L
O P E D   A S P S F O R M E
M U S   I T S B E S T   O U R
E S O T E R I C A   I L L S
    L O S E   A R M L E T
E T U N   K N E E C A P
D E T   E S T A T E T A X E S
D R I E D   E D A S   L D R S
I R O N Y   S I T I   E I N E
E I N E S   T E S T   S N I T
```

PUZZLE SOLUTION 59

```
B E A T   E G G A R   C A W S
A T L I   A T T W O   O D B O
T H E G A R D E N O F E D E N
E N G R A M   S O U N D S
    E A U S   B T O R
H A I   A F C   E S D   T Y S
E R N I   F A R R   I C O M E
D E N N I S T H E M E N A C E
G W Y N N   M E A N   N I A S
E E C   T E A   V E N   R S T
    Y A N N   E M T S
C R O A K S   O S T E N D
T H E P E L I C A N B R I E F
R U N A   E C O L I   I N S C
L E S T   R U N I C   P E I S
```

PUZZLE SOLUTION 60

```
D I S C   O P E D S   N A O H
R E M O   N O R A H   A N T E
I R A N   L O N G E S T D A Y
B I S C A Y N E   Q U A Y S
    E R I A   A Z U R
A R F A R F   A C T A E O N
G A O L S   A B E E T   R E Y
A N N S   A L A N W   S E A R
L C D   O V E T T   A T O L L
H A Y R I D E   O C A S E Y
    S O S A   F M A N
S P A H N   U R E T H A N E
K A R A O K E B A R   O L I N
I S U P   L I E U T   P A C E
M T G E   U S R D A   E E K S
```

PUZZLE SOLUTION 61

```
E M E R   B L Y S   I P S E
N A S A   L I E A T   N A E S
V I S I T I T S W E B S I T E
  N I N E V E H   N O T O F
    R A E R   H O W E
A B O A R D S H I P   A L T A
T E R I S   I L A   D A H L
T A O   M R H Y D   B E C
A T U B   A K A   D B A S E
R I T A   N O T A C H A N C E
    L E T S   R I O S
  A R O N I   L I S T E N S
I C A N T S A Y A S I H A V E
G A P E   S U N N Y   I M E T
A R T Y   A F X E   T U N S
```

PUZZLE SOLUTION 62

```
C I H A T   S A A B S   R S T
O G A G E   A L L A H   A P O
F U N E R A L D I R E C T O R
F A D   N A A   E R R A T U M
E N O L   B M A N   R A I S E
E A U C L A I R E W I   E E N
  S T D S   Y E W   E R S T
  T T T T   S I T E
I L U V   V E E   B R U N
S E N   I S A A C A S I M O V
R T R E V   R U D Y   E B R O
A S A R A I L   R E O   R U T
E N T E N T E C O R D I A L E
L O E   K A S E M   I N G E R
I T D   A S S T S   C H E S S
```

PUZZLE SOLUTION 63

```
E L H I   K L I N K   A G U A
L E T T   A A R O N   P A S T
F A M O U S R A C E H O R S E
S P L O S H   S K E E T E R S
    K E M P   S L S
W O P   R I T A   S T R E A M
I T L L   R R U N   I O N I A
L O U I S I A N A P A R I S H
E M M E T   P T U I   T A L A
D I A D E M   S R O S   C E L
    L A O   U N L V
L A U D A N U M   E U R O P E
S E C U R I T Y D E P O S I T
A N S E   L E D E R   O S S A
T A D S   A R C O S   M O H S
```

PUZZLE SOLUTION 64

```
B A S S   Q U A K E   S I D E
R I T E   U L C E R   E N I D
O N E A N O T H E R   A C E D
S U M M I T   E N O R M I T Y
    A S A P   R U E D
I N A N E   O P T   D R E S S
B I B L I O P H I L E   N T H
E E R Y   O L O G Y   S T O A
A C E   T H I T H E R W A R D
M E A N S   N O T   H A L E Y
    C I A O   S P E D
A S T E R I S K   L A D D I E
C H I C   L O N G I S L A N D
M A N E   E R O S E   E D D A
E G G S   D E B A R   D O O M
```

PUZZLE SOLUTION 65

S	E	P	S		A	L	O	N	E		I	F	A	T
T	R	O	I		C	E	L	E	B		N	I	G	H
A	L	U	M		C	H	A	I	N	S	T	O	R	E
R	E	F	I	N	E	R	S			A	E	R	E	O
		L	O	D	E		D	O	G	G	I	E	S	
F	O	R	E	V	E	R	E	M	B	E	R			
G	I	N	S	U			F	O	O		A	T	C	O
O	S	A		M	A	N	I	P	L	E		R	E	F
Z	E	S	T		N	O	L		L	O	U	I	S	
		C	O	N	V	E	Y	O	R	B	E	L	T	
E	S	O	B	E	S	O		A	N	O	A			
S	C	H	O	N			O	K	E	Y	D	O	K	E
T	A	G	Y	O	U	R	E	I	T		I	N	E	S
O	R	E	L		P	R	I	M	O		A	N	N	I
S	P	E	E		A	S	L	A	N		H	A	L	T

PUZZLE SOLUTION 66

C	A	L	E	B		N	I	H		B	C	E	L	L
A	B	A	C	A		A	D	E		L	O	B	E	D
S	A	C	A	G	A	W	E	A	D	O	L	L	A	R
H	D	I	R	J	B		O	R	U			A	F	S
		T	O	E	S		S	C	I	S				
	I	V	E	B	E	E	N	T	A	K	E	N	I	N
G	H	I			T	E	A		L	O	L	I	T	A
L	A	D	D	S		L	I	S		N	L	E	R	S
O	T	O	O	L	E		F	H	S		C	I	T	
P	E	R	S	E	U	S	S	U	E	S	R	E	P	
			E	R	B	E		I	R	A	E			
B	H	A			I	L	O		I	L	L	S	A	Y
D	O	N	T	S	E	E	E	Y	E	T	O	E	Y	E
A	M	I	R	S		C	N	S		E	S	T	E	E
Y	E	L	L	S		T	O	L		N	E	A	R	S

PUZZLE SOLUTION 67

M	A	L	A		T	R	I	B		I	W	A	L	K
O	L	A	S		H	O	P	I		R	E	N	E	E
H	E	P	C		E	M	A	G		A	D	N	A	N
M	A	T	I	S	S	E		T	I	N	I	E	S	T
			I	N	A		B	O	V	I	D			
I	M	B		A	U	D	R	E	Y		N	N	W	
F	R	O	U	F	R	O	U		L	E	O	I	I	
A	L	L	S	U	I	T	S	S	L	A	S	H	E	D
L	E	O	N	S		Q	U	A	R	T	I	L	E	
L	E	S		S	T	U	P	O	R		T	S	R	
		H	O	R	A	E		B	Y	A				
O	U	T	S	W	I	M		T	V	S	T	A	R	S
A	B	R	I	L		A	N	U	N		T	V	I	L
K	R	E	M	E		R	E	B	S		A	I	R	E
S	A	Y	I	T		A	C	A	P		R	V	E	R

PUZZLE SOLUTION 68

O	N	E	L		A	D	I	E	U		H	G	T	S
R	A	S	E		L	A	L	M	P		U	P	O	N
R	U	S	T	B	U	C	K	E	T		B	S	E	A
S	T	U		A	M	E	S		O	R	B			
			I	B	N			H	A	U	T	E	S	
T	R	A	C	Y	A	N	D	H	E	P	B	U	R	N
K	A	S	E	M		O	R	A	R	E		T	V	A
T	I	A	S		T	W	I	N	E		S	T	I	R
O	N	S		E	R	I	E	S		A	T	U	N	E
C	E	E	E	N	E	N	R	E	P	O	R	T	E	R
K	S	T	A	R	S			O	K	S				
			G	Y	P		S	E	T	A		J	I	A
W	A	L	L		A	L	L	D	A	Y	L	O	N	G
M	U	S	E		S	O	I	N	G		T	I	R	E
P	S	T	S		S	E	T	A	E		D	E	E	R

PUZZLE SOLUTION 69

M	A	P	H		S	N	C	C		H	A	S	T	E
K	L	E	E		T	O	O	L		E	N	E	S	T
T	O	O	G	O	O	D	T	O	B	E	T	R	U	E
G	E	N	I	E			S	U	L		B	T	P	S
			R	I	G	A		D	U	P	E			
L	O	C	A	L	P	H	O	N	E	C	A	L	L	
E	L	A	S		A	S	H			B	R	A	E	S
A	L	I		I	S	O	B	A	R	S		I	I	N
D	A	N	D	D			O	N	A		A	N	T	I
	S	E	R	V	E	S	Y	O	U	R	I	G	H	T
			A	E	R	I		A	L	E	R			
A	B	R	I		O	E	O			P	S	S	S	T
I	M	I	N	T	O	S	O	M	E	T	H	I	N	G
R	O	N	E	E		T	Z	A	R		O	N	E	L
E	C	A	R	D		A	E	N	A		W	E	R	E

PUZZLE SOLUTION 70

O	N	K	P			I	F	A		M	E	A	R	A
C	A	S	O	S		S	U	I		E	D	G	E	S
H	M	E	E	C		L	B	S		O	B	I	T	S
O	U	T	T	O	L	A	U	N	C	H		R	I	E
			E	L	E	M		E	O	M	E	L	E	T
M	O	B		A	R	A	M		A	Y	N			
K	N	O	R	R		B	O	A	C		T	E	R	I
T	O	S	S	I	N	A	N	D	T	U	R	N	I	N
G	R	N	I		E	D	D	A		N	E	R	D	S
			D	A	H		E	M	I	L		Y	E	T
T	H	R	E	E	R	S		A	R	E	E			
E	O	E		S	U	S	A	N	T	A	N	G	O	S
L	O	T	T	O		T	U	T		S	T	A	R	T
I	C	E	U	P		A	L	L		H	S	I	M	I
C	H	R	I	S		R	D	Y			E	N	E	R

PUZZLE SOLUTION 71

E	M	B		Q	U	E	U	E		S	K	I	M	S
S	O	O		U	N	A	R	M		E	U	R	O	S
T	U	R	B	O	P	R	O	P		I	L	E	N	E
A	L	A	R	I	C		A	I	N	T				
D	I	G	A	T		S	S	T	S		U	N	U	M
O	N	E	S		U	P	T	H	E	C	R	E	E	K
		S	O	R	E	E	Y	E	S		I	N	T	
N	U	T	M	E	G			Y	I	E	L	D	S	
I	D	I		N	E	A	R	M	O	S	T			
C	O	R	D	O	N	B	L	E	U		T	S	C	A
K	N	O	T		C	U	S	S		T	I	E	O	N
		R	E	Y	S			E	R	N	A	N	I	
E	L	I	A	N		I	N	E	L	E	G	A	N	T
V	I	S	I	T		V	E	N	U	S		I	E	R
S	T	E	N	O		E	N	C	L	S		R	D	A

PUZZLE SOLUTION 72

E	A	R	O	F		M	A	T	T	S			S	V	E
L	L	O	S	A		O	N	R	I	O			H	A	B
E	L	S	I	E		I	G	O	T	A	N	A	M	E	
M	Y	H	E	R	O		E	P	I		O	N	O	R	
			O	V	U	L	E		E	G	E	S	T		
A	D	I	M	E	A	D	O	Z	E	N					
T	E	M	A		T	O	S		A	R	A	R	A	T	
W	A	P	I	T	I	S		P	S	Y	C	H	U	P	
T	E	S	O	R	O		E	N	E		R	I	T	E	
		O	N	O	N	E	S	H	O	N	O	R			
L	A	L	M	P		D	N	U	O	P					
O	D	I	U		E	D	O		N	O	H	E	L	P	
I	N	M	Y	T	R	I	B	E		W	I	Z	E	N	
R	A	B		T	I	T	L	E		E	V	E	N	I	
E	N	S		L	A	Y	E	R		R	E	R	A	N	

PUZZLE SOLUTION 73

PUZZLE SOLUTION 74

PUZZLE SOLUTION 75

PUZZLE SOLUTION 76

PUZZLE SOLUTION 77

```
C H A R R O   P L E D   O O O
R I T E O F   L E D E   K W A
E T H A N E   A N G R I E S T
A S O D D   S Y N E R G Y
T O M D O O L E Y   O D E D
E N E S   A A R   K E T O N E
      M K T S   C R I K E Y
E D G A R L E E M A S T E R S
C H I G O E   N Y S E
C O L O N Y   T R E   R I D I
O W E R   C R O S S W O R D
  V A U G H A N   H A L A L
T E A S P O O N   M E N A G E
M T N   U N P C   R A D N O R
S H S   P O S E   B R A I N S
```

PUZZLE SOLUTION 78

```
W E L C O M E M A T   I V O R
I N A C T I V A T E   W E R E
C O N S O N A N T S   A S I S
C U D   H I S N   I S P E P
A G R A   A B A M   U N O
  H U S H U P   A T P   C T N
  S E L E C T O R   C A S
C L I E N T S C A S E F I L E
H E N   C R E V A S S E
E C H   O A T   N A S D A Q
E T A   O S A Y   S E U L
T U L I P   A A R P   N A A
A R A L   H E R B A L P E R T
H E N S   A N N I H I L A T E
S R T A   P O S T S E A S O N
```

PUZZLE SOLUTION 79

```
M E A T Y   S T O N   S M U G
E N N I O   C E L O   H O A D
A D I E U   P R E T M E N T S
G E O R G I E P O R G I E
R A N S O M   A S K Y O U
E R S   G O A D E D   P O R
  K I N G S M E N T I O N
C A B E R   G T I   E S T H S
S M E L L A R A T H E R
I B A   H O R S E D   S N A
S I R E N E   M I S T E R
  D A M A G E C O N T R O L
E L I M I D A T E   E R A S E
N E N E   O R D O   S E T H E
C I G S   F E S S   S W O O N
```

PUZZLE SOLUTION 80

```
B R A W L   S C H M O   E F S
R E T A X   C R A E S   M O T
O B E L I   H O M E S   O R K
W O R K I N W O R K O U T S
N O G O   K A K A   R I A L
S T O V E R S   D S T I C K S
  E M U   C I I I   O E D
  J E R O M E R O B B I N S
H A R   T A X A   E I N
M I N N E H A   B R A L E S S
M L I I   M A U I   A L O U
  B E N E D I C T A R N O L D
E A L   R E N A L   A D I O S
S I L   R E E S E   M E S N E
S T L   S P D E R   P R E S S
```

PUZZLE SOLUTION 81

PUZZLE SOLUTION 82

PUZZLE SOLUTION 83

PUZZLE SOLUTION 84

PUZZLE SOLUTION 85

```
L O T   L A P S E   B A F T A
O N A   A B R A M   E L A N D
N E K   M A O R I   D A N D D
G A E L I C W A R N I N G
T R A V A I L     E M I L I E
O M N I   S I C A   S I L O
    P R I   N L R B   K E N
    I G O T Y O U B A B E
R Y N   E S A I   Y R S
K A L I   Y O L K   I N X S
O P E N T O   I N A D A Z E
    A D J U S T S T H E M I R
D C G U Q   P A S E O   A B U
S H U C K   A L I S O   T I M
M E E T A   R I N T T   H T S
```

PUZZLE SOLUTION 86

```
S A D O   A D I A     U N A
A M O F   S O D S   S Y N C S
D O G F I S H E S   M A D A S
R I S E T O   O N S A L E A T
    R O C K   T N U T
E M L E N   A S R E D   E A U
Q A I D   N B A E R S   R B S
U T E   W O O L E N S   R E E
I S A   E R O I C A   P E E N
P U B   E E M C T   A D D T O
  I A L E   S U B J
B E L C A N T O   T E A R A T
E R I N S   N A K E D M A J A
A N T E S   P H O N   E N A M
M A Y   K U H N   S I M S
```

PUZZLE SOLUTION 87

```
D R E A   C B G B   E T A P E
W E L L   R I L E   N H L E R
A B A A   I N O R   I I I I I
R U N N I N G P R E S S E S
F T D I X     Y E L L
    S N A G S   W E I R D O
A M P   A V O E R   F E A R
S O R R Y I S P I L L E D I T
C R E E   S T A V E   S S H
H K P D C Y   I L I U M
    R A E R   M E S S Y
  Y M I G H T B E G I A N T S
D R O V E   R A I L   L E I A
O L D E R   E L L A   I L L Y
A Y E R S   V L A D   E L E E
```

PUZZLE SOLUTION 88

```
A H A S   W A A C   B S I D E
C O L I   T A L I   A P R O N
E R I N M O R A N   T H E O C
H A S T A   P E C U N I A R Y
    A R A   F E N D S
X X X X X X X X X X X X X
L I K E   L A T E S T   D P I
V I E S   X I N   H E R N
I I S   E R E N O W   E P I C
  T H E S A N D P I P E R
  E T S E Q   S C M
L A S E R D I S K   B L I G E
E M A T E   T R A N S I E N T
V E R S A   S I E G   C R U E
I S S E L   I S L S   H E S S
```

PUZZLE SOLUTION 89

S	I	T	A		S	C	O	W			M	B	E	K	I
T	A	R	N		I	A	G	O			E	U	R	O	S
A	L	E	E		N	L	E	R			D	R	I	L	L
B	L	A	C	K	E	Y	E	D	S	U	S	A	N		
			D	R	A	X			S	L	A				
M	R	T	O	A	D		N	A	S	L		S	P	H	
I	E	A	T			S	O	U	S	A		T	I	A	
T	R	U	E	C	O	N	F	E	S	S	I	O	N	S	
E	I	N		A	R	O	A	R			N	I	S	I	
S	G	T		N	E	W	T		B	L	A	C	K	D	
			E	D	O			G	U	A	C				
	P	O	L	I	S	H	V	A	C	A	T	I	O	N	
C	O	U	L	D		T	O	R	K		I	R	M	A	
U	N	T	I	L		T	I	D	E		V	A	R	Y	
B	Y	A	N	Y		P	R	E	T		E	N	I	S	

PUZZLE SOLUTION 90

U	N	A	M	I		J	I	B	S		L	A	I	T
R	E	R	I	G		O	N	E	A		I	T	S	O
O	U	N	C	E		I	G	A	S		E	L	A	T
	F	O	R	T	Y	S	E	V	E	N	D	A	Y	S
		O	N	A	T				O	T	W	A	Y	
B	A	M	B	O	O		D	A	R	I	O			
I	D	A	E		S	E	R	A	C		C	I	E	
T	O	P	S	P	I	N	F	O	R	E	H	A	N	D
O	S	H		A	G	O	O	D		A	G	C	Y	
		N	I	O	B	E		E	D	S	E	L	S	
S	A	P	O	R		A	L	O	F					
E	X	C	U	S	E	M	E	P	L	E	A	S	E	
M	E	A	G		R	O	E	S		S	I	S	A	L
I	L	K	A		E	E	R	O		N	T	E	S	T
S	S	E	T		S	S	S	S		T	H	E	Y	S

PUZZLE SOLUTION 91

U	N	T	O		R	A	K	E	D		S	A	N	A
S	C	A	S		I	N	U	R	E		T	R	I	S
C	A	S	E	O	F	T	H	E	C	R	E	E	P	S
G	A	M	E	P	L	A	N		I	O	N	I	A	N
			C	E	E		I	M	S	O				
	A	D	D	I	S		N	C	A	A	G	A	M	E
K	R	A	I	T		G	O	A	T		P	O	D	
T	A	K	E	S	O	N	T	H	E	C	H	A	I	N
E	D	A		D	A	R	N		C	A	R	L	A	
L	O	R	D	H	O	W	E		R	C	P	T	S	
		E	E	R	S		E	E	L					
A	R	F	A	R	F		S	A	T	I	R	I	S	T
C	O	N	S	E	R	V	A	T	I	V	E	T	I	E
R	U	M	I		E	A	M	O	N		P	I	C	A
E	T	A	L		E	L	I	N	A		O	N	A	T

PUZZLE SOLUTION 92

A	N	D	I		A	P	P	T		S	P	O	O	N
R	O	I	S		D	E	A	R		H	A	V	R	E
I	T	S	D	E	J	A	V	U		U	S	E	M	E
B	A	R	N	A	C	L	E	C	H	E	S	T	E	D
R	O	A	S	T		M	E	E		I	T	R	Y	
		B	O	T		D	E	S	E	R	T			
M	D	I		A	N	T		H	O	B	B	Y		
G	E	N	R	E		S	T	O		E	N	E	S	T
M	C	G	E	E		K	A	R		A	S	D		
	D	R	E	A	R	Y		D	A	U				
L	A	D	I		M	P	T		M	O	R	T	E	
T	H	E	V	I	I	I	I	I	I	I	I	I	N	E
G	E	L	I	D		T	S	E	T	S	E	F	L	Y
E	R	O	D	E		A	T	A	T		G	U	A	R
N	O	N	E	S		L	S	T	S		E	L	I	E

PUZZLE SOLUTION 93

```
O D R A ▓ C D G E L ▓ E G A D
S C O P ▓ E R A T O ▓ G A E A
S U M O ▓ D U L A C ▓ G E R M
A P A S S A G E T O I N D I A
▓ ▓ T I R ▓ ▓ W Y O ▓ ▓ ▓ ▓ ▓
R E A L E S T A T E A G E N T
T E L E G ▓ O M B E R ▓ M O R
R N A S ▓ S T O O D ▓ A B I E
E I S ▓ A T O L L ▓ P L O C E
V E T L E T P E T G E T W E T
▓ ▓ O O H ▓ ▓ ▓ O R E ▓ ▓ ▓ ▓
F I L M N O I R A C T R E S S
R T E E ▓ M O T T O ▓ A D E E
E R N I ▓ A N E E L ▓ N E X T
D Y A N ▓ S O S A D ▓ T R Y A
```

PUZZLE SOLUTION 94

```
A D A M ▓ H T T M ▓ ▓ T A E M
C A M P H O R O I L ▓ A N G E
H E I G H T O F F A S H I O N
E W E ▓ H E P ▓ ▓ P E R M I T
N O N E ▓ P I R A T E ▓ A S I
E O S I N ▓ C E L ▓ ▓ A T T O
▓ ▓ ▓ G O T ▓ B A L L P E E N
▓ C H O W D E R H E A D ▓ ▓ ▓
B R A T P A C K ▓ Z I T ▓ ▓ ▓
A I R H ▓ ▓ C A T ▓ F H O L E
H O P ▓ P E L H A M ▓ Y G O R
A L U M A E ▓ R I A ▓ L O N
M O T I O N S T O S T R I K E
A B E T ▓ S T A T E M E N T S
S O R E ▓ ▓ U S S R ▓ I G O T
```

PUZZLE SOLUTION 95

```
▓ M C D ▓ S E R I N ▓ T C I
H E I R S ▓ U S E N O ▓ H O B
T A C O T O P P I N G ▓ E N E
E L E V E S ▓ S N A G G I N G
S I L E N T B ▓ A R I E S ▓
T E Y ▓ T E E M ▓ D N A L A B
▓ ▓ W O R K T F S ▓ N E R A
R U B E N ▓ A W E ▓ R I Y A L
U P A S ▓ B A T T E A U
P I S T O L ▓ F I L I ▓ H S N
▓ E I R E A ▓ D O N A T E S
E P H E M E R A ▓ P R I O R I
T A I ▓ O P E N H E A R T E D
T K T ▓ L E C I D ▓ I S O N E
A S S ▓ U D A L L ▓ ▓ E E E
```

PUZZLE SOLUTION 96

```
A L P S ▓ H S I A ▓ ▓ S U L A
B O U T ▓ A O R T A ▓ E G I S
R A S P ▓ M O E S Z Y S L A K
I M M E R S E D ▓ E O S I N E
▓ ▓ O T W A Y ▓ C R U I S E R
S H T E T L ▓ C A I R O ▓ ▓
I A T ▓ J A V A N ▓ E N F C E
S H O T ▓ D E R E K ▓ S L A M
I N B I G ▓ R E S O D ▓ A L I
▓ ▓ P U S S Y ▓ K I S S I N
G O D P A P A ▓ D O E T H ▓
E L A Y N E ▓ G I M M I C K Y
T I N T O R E T T O ▓ G A L E
N O G O ▓ M B E K I ▓ M R E S
O S S E ▓ N A A N ▓ A D E S
```

PUZZLE SOLUTION 97

```
T G I   O T O S E   N L E R
H E N S   P E R I T   A A R E
U N S U I T A B L E   U T I L
G E T S S O R E   R I S E T O
S T R A T U S   M N D E L
    N I T   L A I L A A L I
F R A N C   B E R T S   T A M
L U R E   N A G G Y   R I S A
A E C   T E R S E   N E N E S
M R W I Z A R D   B E V
    E M A T E   L E M O N D E
V O L A R E   D A T E L E S S
I H D R   N R A D I A T I O N
R T E E   U A N D D   S L U E
T O R T   P U L S E   L P S
```

PUZZLE SOLUTION 98

```
C O S A   B M A J   F A C E R
A R U T   I E R I   A L A D Y
R A Z E   E N A M   N I S A N
T R Y S U N G L A S S E S
S E Q U I N S   T I N E A R
    P E I   C E R T   T Z E
I N S P   A L E X K E A T O N
V A L E T   U N E   S M E L T
A B E R R A N T L Y   E S E S
N E E   I T E R   E U R
A S P E C T   P A R I A H S
    G R O U N D C H U C K L E
M O B I L   I Y A M   I E A T
E R A T O   K A K A   U L N A
A U G U R   I D E N   M A D E
```

PUZZLE SOLUTION 99

```
A A B A   C A N A P   O O N A
L R O N   E R A O F   A N O D
L A B O R D A Y W E E K E N D
H Y S T E R I A   N A T H A N
    H A I D   A N D A
K F S E L C   A V I   G R I A
A L E R T   U S A G E   T R S
N O T S O L D I N S T O R E S
I O S   R E A C T   E V E N I
N D A K   A L E   L S E V E N
    A D D L   V A I R
O C T R O I   D I G A H O L E
G O V E R N O R G E N E R A L
L I M N   T E N O R   R R N A
E R A S   O D O R S   E S E S
```

PUZZLE SOLUTION 100

```
A B O M A   I N L A   L T R
E L O R O   C L I P S   E E E
C O L O R F U L A N I M A L S
  C O N T O R T   O N N O
    I X I A   E M C E E D
A R T I C H O K E H E A R T S
R E E D   E N E S
R O S E M A R Y C L O O N E Y
    O F N O   I M R E
A T O M I C S U B M A R I N E
T A B U L A   T T O P
A M E S   H A T R A C K
M A Y T A G R E P A I R M A N
A L E   H A V R E   L A I N G
N E R   N Y S E   S P I E S
```